D1453240

ECLIPSE OF MAN

NEW ATLANTIS BOOKS

Adam Keiper, Series Editor

PREVIOUS VOLUMES:

Why Place Matters:
Geography, Identity, and Civic Life in Modern America
Edited by Wilfred M. McClay and Ted V. McAllister

Merchants of Despair:
Radical Environmentalists, Criminal Pseudo-Scientists, and the Fatal Cult of Antihumanism
Robert Zubrin

Neither Beast nor God:
The Dignity of the Human Person
Gilbert Meilaender

Imagining the Future:
Science and American Democracy
Yuval Levin

In the Shadow of Progress:
Being Human in the Age of Technology
Eric Cohen

www.newatlantisbooks.com

Charles T. Rubin

Eclipse of Man

Human Extinction and the Meaning of Progress

New Atlantis Books
Encounter Books · NEW YORK · LONDON

First American edition published in 2014 by Encounter Books, an activity of Encounter for Culture and Education, Inc., a nonprofit, tax exempt corporation.
Encounter Books website address: www.encounterbooks.com

Manufactured in the United States and printed on acid-free paper. The paper used in this publication meets the minimum requirements of ANSI / NISO Z39.48-1992 (R 1997) (*Permanence of Paper*).

FIRST AMERICAN EDITION

LIBRARY OF CONGRESS CATALOGING-IN-PUBLICATION DATA

Rubin, Charles T.
Eclipse of man : human extinction and the meaning of progress / by Charles T. Rubin.
pages cm
Includes bibliographical references and index.
ISBN 978-1-59403-736-8 (hardcover : alk. paper) —
ISBN 978-1-59403-741-2 (ebook)
1. Philosophical anthropology. 2. Human beings—Forecasting. 3. Human evolution. 4. Human body—Technological innovations. 5. Cyborgs. 6. Biotechnology—Moral and ethical aspects. 7. Humanity. 8. Progress. I. Title.
BD450.R73 2014
128—dc23
2014022022

CONTENTS

FOR MY PARENTS

I know that it is a hopeless undertaking to debate about fundamental value judgments. For instance, if someone approves, as a goal, the extirpation of the human race from the earth, one cannot refute such a viewpoint on rational grounds. But if there is agreement on certain goals and values, one can argue rationally about the means by which these objectives may be attained.

ALBERT EINSTEIN, 1940

Introduction

"MANKIND WILL surely destroy itself." Whether predicted in a thunderous denunciation of our flaws or with mild worldly regret, an apocalyptic future has become a cliché. Will it be global warming or global cooling? Nuclear winter or radiation poisoning? Famine due to overpopulation, or pollution-induced sterility? These are some of the possibilities I grew up with. But today, it is becoming increasingly common to hear of another route to the demise of humanity: we will improve ourselves, becoming something new and better, and in doing so we will destroy what we are now. We have this opportunity because science and technology are giving us the power to control human evolution, turning it from a natural process based on chance to one guided by our own intelligence and will.

This idea—that human progress points toward human extinction—is held by people who go by a variety of names: transhumanists, posthumanists, extropians, advocates of H+, or singularitarians. It can be difficult to keep these terms straight, as they each represent schools of thought whose agreements and disagreements can be complex and ingrown. For the purposes of this book, all these schools of thought will be given the generic label *transhumanism*.[1] The essential insight that defines transhumanism is, to borrow a phrase, that "We are as gods and might as well get good at it."[2] Transhumanists argue not only that modern science and technology are giving human beings the power to take evolution into our own hands to improve the human species, and then to create some new species entirely, but also the ability to improve on all of nature. Much like the older apocalyptic visions, the transhumanists believe that mankind as we know it and nature as we know it are on their way out; but for most transhumanists, that is the deliberate goal sought, not a consequence of our hubris to be avoided. Indeed, the transhumanists believe that if we are to prevent some of the more common apocalyptic visions from becoming reality, we *must* redesign humanity so that our ruinous flaws can be eliminated. To avoid mere destruction, we must embrace creative destruction.[3]

Hence, the end of man can be the beginning of . . . who knows what, exactly? But, we are told, it will doubtless be some wondrous new home for intelligence, able to do things far beyond our present ability to imagine. If my baby boom generation was warned that the challenge of the future involved ensuring the continued existence of human beings in the face of all the threats posed by our own activities (and to some extent by nature itself), for the transhumanists it is becoming increasingly difficult to imagine a future that has any place for us at all, except perhaps as curiosities.

Such views of the future can strike sensible people as idiosyncratic or even loony. But this book will show why that understandable first reaction should not be the whole story. There are some serious reasons why transhumanists have come to measure progress by the speed at which mankind disappears, and those reasons are more deeply rooted in mainstream ways of looking at the world than might at first be obvious. At the same time, as we shall see, there are also very good reasons to reject the transhumanist future, and to work toward a future in which, as William Faulkner put it in his speech accepting the Nobel Prize in Literature, "man will not merely endure: he will prevail."[4] Those reasons will emerge from a serious confrontation with transhumanist ideals and the foundations on which they are built.

SUFFICIENT UNTO THE DAY?

It can be difficult to know what transhumanism amounts to—a worldview, an ideology, a movement, or some combination—but it cannot simply be dismissed as irrelevant. Doctoral dissertations and academic conferences have focused on transhumanism, and a few major universities have scholarly centers wholly devoted to exploring transhumanist ideas. Books and blogs, think tanks and online communities, documentaries and blockbuster movies have all helped to popularize those ideas. Major news outlets routinely publish reports uncritically explaining them.[5] Meanwhile, some of Silicon Valley's best and brightest are committed transhumanists. Perhaps the most prominent promoter of transhumanism, the inventor and bestselling author Ray Kurzweil, subject of countless media profiles, was hired by Google in 2012 to serve as the company's director of engineering.[6] Three years earlier, Kurzweil cofounded Singularity University, an institution dedicated to disseminating transhumanist ideas, with sponsorship from several high-tech companies and philanthropic foundations, as well as help from NASA.[7] In short, many of the people who are inventing the tools of tomorrow embrace, or are at least informed by, the transhumanist vision of the day after tomorrow.

The transhumanist program to redesign humanity is often linked with the rise of the so-called "converging technologies": nanotechnology, biotechnology, information technology (and sometimes robotics), and cognitive science.[8] They are called converging technologies because each reinforces the potential that the others have for vastly increasing our ability to manipulate nature, including our own nature. And what remarkable things are now in the works! While it is still possible to make a splash by writing a book about how, in the not-so-distant future, people will regularly have sex with robots,[9] that is hardly a revolutionary thought when for some time others have been writing books about people turning themselves into robots.[10] The "virtual reality" expected soon to make movies and games more immersive is just a precursor to direct connections between our brains and computers, and even that is merely a prelude to uploading our minds into computers—providing us with a kind of immortality (so long as proper backups are made).[11] The undoubted promise biotechnology holds for lengthening human lives is overshadowed by speculation concerning the ability of nanotechnology to bring the dead back to life.[12] We are told that genetic engineering to cure disease and bioengineering to overcome disabilities are just foreshadowings of a complete re-engineering of human beings to add whatever senses, features, and capacities an individual might wish to possess.[13]

Are such developments really likely, or even possible? Some critics point out problematic scientific or technological assumptions underlying transhumanists' ideas, or the obstacles that they might have overlooked. Critics also sometimes focus on the considerable uncertainty about what is possible, arguing that that uncertainty might in and of itself be adequate reason for ignoring those who happily anticipate an end to humanity. Without a clearer idea of what will actually be possible in the future, debating the details of various transhumanist predictions might seem like a waste of time—and surely we have more pressing things to worry about today.

But even if the converging technologies do not pan out exactly as transhumanists expect, who would really want to bet against the likelihood that science and technology will, in the future as in the past, continue to allow us to do things routinely that only a few decades previously might have seemed like mere science fiction? Indeed, it is surely likely that for a great many items on those worry-about-it-today lists, we will call upon science and technology to deal with them. That is our contemporary way of thinking about solving problems. Nanotechnologies are already being developed to deal with environmental and energy issues.[14] The fight against terrorism has spurred advances in robotics.[15] The most surprising human future would be one in which we did *not* continue to accumulate scientific information and innovative technology and use them to increase our powers over nature, which includes power over ourselves.

From this point of view, it is clear that the broad transhumanist goal of overcoming the human condition does not depend on whether or not a *particular* constellation of technologies works as expected. Indeed, we will see how some transhumanists reasonably suggest that dissatisfaction with the human condition and the wish to transcend it run deep in human thinking, and perhaps even define our humanity. If so, we may just keep trying to redesign ourselves with whatever means actually become available.

MAN IS BORN TO TROUBLE

Dissatisfaction with the miseries of human life—whether we are beset by them from outside or we bring them on ourselves—is nothing new. In the Bible, the Book of Job lays out the situation in a familiar way from which certain generic conclusions can be drawn. Job loses his great wealth, the external goods that make for a comfortable life. Boils afflict his body and take his health away. Job is deprived of his loved ones, and to that extent also of his future, as he fears that death is absolutely final. The only thing missing is deliberately omitted. Satan expects the balance of Job's mind will be disturbed and he will curse God. But while Job acknowledges confusion, longs for death, and contends with God, he does not explicitly curse Him.

So the human condition has long been understood to include the possibility, indeed the likelihood, of being deprived of external goods, bodily goods, and goods of the mind or spirit—and by dissatisfaction that we have such vulnerabilities. Doubtless that kind of dissatisfaction helped to prompt ancient imagination of longer lives, greater wealth, and superhuman power, as in the case of the Greek gods and heroes. These Greek gods—without the curse of mortality, with all possibility of ease and wealth and security, full of rude health and bodily vigor—are just what we might wish to be. And yet they are still restless, jealous, capricious, untrustworthy, often angry, unhappy, and quite dissatisfied. Even without the darker sides of the human condition, and without having to bear real consequences for the vast majority of their actions, they have a terrible lightness of being.

While these gods will punish those who seek to be too much like them, the gods themselves are punished for being what they are, for having some of the very goods that mortals hope for. Apparently, in the ancient view, the things we might have thought would make us happy do not guarantee it at all. Failure to appreciate this catch-22 means that imagination of something better is likely to be just another source of suffering, since what you imagine would never work out in any case. So whether you think things could be better or not, suffering is to be taken for granted. Options for dealing with this fact of life, whether among premodern polytheists or monotheists, were likewise limited. One might seek to delay suffering by proper relationships with gods or God, or to find some message or meaning in the suffering when it inevitably occurred, or simply to accept it with resignation as the cost of living. And at that, even the pious rabbis of the Talmud were known to wonder if it were better for mankind to have been created or not. When it came to a vote, they voted not.[16]

Even if the sources of our misery have not changed over time, the way we *think* about them has certainly changed between the ancient world and ours. What was once simply a fact of life to which we could only resign ourselves has become for us a problem to be solved. When and why the ancient outlook began to go into eclipse in the West is something scholars love to discuss, but that a fundamental change has occurred seems undeniable. Somewhere along the line, with thinkers like Francis Bacon and René Descartes playing a major role, people began to believe that misery, poverty, illness, and even death itself were not permanent facts of life that link us to the transcendent but rather challenges to our ingenuity in the here and now. And that outlook has had marvelous success where it has taken hold, allowing more people to live longer, wealthier, and healthier lives than ever before.

So the transhumanists are correct to point out that the desire to alter the human condition runs deep in us, and that attempts to alter it have a long history. But even starting from our perennial dissatisfaction, and from our ever-growing power to do something about the causes of our dissatisfaction, it is not obvious how we get from seeking to *improve* the prospects for human flourishing to *rejecting* our humanity altogether. If the former impulse is philanthropic, is the latter not obviously misanthropic? Do we want to look forward to a future where man is absent, to make that goal our normative vision of how we would like the world to be?

DREAMS OF THE FUTURE AS MORAL VISIONS

My previous book culminated in a discussion of deep ecology, a form of radical environmentalism that adopts the principle of "eco-egalitarianism": no one species has any moral priority over another. Because technological man so consistently violates this stricture, some of the deep ecologists look forward to a day when human beings will have been replaced by a new human-like species that lives more at one with nature. I criticized that view for its profoundly anti-human character, but even then I noted that the same threat could come from the direction of technological utopianism.[17] Deliberately seeking our own extinction represents the extreme limit of how far we could want to go to overcome our given circumstances and raises in an obvious way a question that is always lurking in our rapidly changing world: What kind of future are we trying to create?

It is very likely that the world we will have in the future will not be *exactly* the one laid out by today's transhumanists. Still, our utter dependence on continuing scientific and technological development makes it impossible to dismiss the broad goals of transhumanism outright; indeed, it is hard to imagine how we will avoid making choices that could provide building blocks for a project of human extinction. Even if in most instances these choices will actually be made with a view to the contingencies of the moment—arising from scientific curiosity, engineering creativity, military necessity, or commercial possibility—the transhumanist grand vision of the eclipse of man will be there to influence, rationalize, and justify favoring certain alternatives. It provides a narrative that takes those alternatives beyond contingency and presents them in a way that intentionally creates dissatisfaction with any merely human account of how we live and treat each other now. If it is the only story going, it is all the more likely to provide the moral meaning behind the scientific and technological future.[18]

An argument could be made that we should avoid taking too seriously such grand visions of the future. In his book *In the Shadow of Progress*, Eric Cohen warns, speaking specifically of genetics, that in order "to think clearly" and avoid the twin vices of over-prediction (assuming that our worst fears or greatest hopes will come to pass) and under-prediction (failure to acknowledge where present developments might go), "we must put aside the grand dreams and great nightmares of the genetic future to consider the moral meaning of the genetic present," instead exploring "what these new genetic possibilities might mean for how we live, what we value, and how we treat one another."[19]

Cohen's caution is well taken, and the questions he poses ought indeed to be where thoughtful people begin to confront the apparently ceaseless innovations of our technological society. But it may be harder than it looks to separate how we think about the present from how we think about the future. Whether in secular or religious terms, it is not unusual for people to define "the moral meaning of the present" in terms of the future, judging what *is* against what they hope (or fear) *will be*. After all, it is moral choice in the present that creates a future, so such visions influence how we live and treat each other today.

Cohen is doubtless right that a more sober and serious moral world would have less place for "grand dreams and great nightmares." Yet given that we cannot but be influenced by such visions, we must come to grips with them on their own terms; there have to be *reasons* for putting them aside. People were not unaware of problems with capitalism before Marx, but Marxism became a sufficiently powerful grand vision that it had just the effect Cohen fears, blinding people to the truth of their present circumstances and of their obligations to those around them. Facing up to the defects of Marxism was easier said than done and remains an incompletely accomplished task. But one part of doing so was intellectual confrontation with it *as a grand vision*. None of the specific transhumanist visions of the future have as yet anything like the intellectual or political power of Marxism at its height. But if we are to eschew them, a similar intellectual confrontation will be necessary.

CONFRONTING THE EXTINCTIONISTS

It can be difficult to find a footing for criticism of transhumanism, given the broad scope of the transhumanist vision, the variety of (sometimes conflicting) transhumanist ideas, and of course the fact that we cannot yet know precisely what direction and shape transhumanist ideas will take in the real world. That is why the focus of this volume will be on transhumanism's *moral* vision of the future, rather than its technological or scientific content. We will approach transhumanism sometimes directly and sometimes obliquely. While we will analyze the ideas of some of the most provocative and controversial transhumanist thinkers, we will also look at a wide range of other materials. We will journey from the invisibly tiny scale of molecular engineering to the far reaches of outer space. We will discuss ancient myths and recent science fiction, centuries-old paintings and Hollywood movies. And we will explore some forgotten byways, learning from long-neglected stories that can teach us about our desires for the future. Several of the chapters will also begin with prologues, short fictional stories I have written to help bring to life some of the ideas addressed in those chapters.

Why the emphasis on fiction? When compared to the works of nonfiction by experts and specialists in transhumanism, very often fictional works present us with a far more realistic and morally nuanced picture of the issues at stake in human self-overcoming. This should not surprise us. After all, excellent fiction requires a serious understanding of human things, and to tell a great story about scientific and technological possibilities, fiction writers have to start from a convincing human world—which of course is the world out of which the scientific and technological developments that will allow human redesign will *actually* emerge. Those who start instead from the imagination of technological possibilities often suffer a kind of tunnel vision, a narrow focus that makes them assume that whatever they are writing about will be the center around which everything else will revolve. Indeed, that outlook helps to account for the moral weakness of the transhumanist project.

When you get beyond transhumanism's fascination with the technological cutting edge, it becomes evident that its hopes are not new. To separate the question of what kind of future is technologically likely from the question of what is a morally desirable future, it is useful to look closely at some of the earlier thinkers who developed ideas that are important to transhumanism—even if the transhumanists do not realize the debt they owe them. While we will not attempt to offer a complete intellectual history of such ideas, in the chapter that follows we will look at a selection of particularly influential presentations of the theme of overcoming humanity, stretching back to the late eighteenth century. As we shall see, these thinkers lay down some of the key foundations upon which today's transhumanists build. We turn there first in order to answer a fundamental question: Why would anyone think that human extinction is a good thing?

CHAPTER ONE

The Future in the Past

THE ASPIRATIONS of transhumanism are not entirely new because the human desires to change unsatisfactory aspects of our lives or to extend our powers are not new. Yet these deeper, perduring aspirations have gained a new force in the modern age, as the world has come increasingly to be defined by the ability of modern science and technology to change it significantly. It has become possible to think seriously about progress, about what it would mean if these achievements were increasingly to shape our lives, building one on another.

Progress does not inevitably have to point to the kind of future transhumanists envision. Science fiction authors, and others who think about the future in less overtly fictional ways, have had no trouble thinking about some pretty remarkable and very distant futures in which humanity persists in an eminently recognizable form. This does not represent a mere failure of imagination on the part of these futurists; in some ways, it arguably takes even more imagination to depict a future in which mankind has survived without becoming radically different.[1]

To understand why the transhumanists believe progress requires human extinction, we can study some of their intellectual forebears—

authors who, while not all widely remembered today, were in their own time influential on the way people understood where science and technology might be taking us. We begin with one of the greatest of Enlightenment prophets of progress, Marie Jean Antoine Nicolas de Caritat, the French aristocrat known as the Marquis de Condorcet (1743–1794).

Condorcet was a mathematician and philosopher, an abolitionist and advocate of women's equality and religious toleration, an admirer and biographer of Voltaire, an economic liberal, and a genial enthusiast—an embodiment, in short, of all things enlightened.[2] A prominent public intellectual, he was "one of the few Enlightenment thinkers to witness the [French] Revolution and to participate fully in its constitutional aftermath. . . . He was, in short, an outstanding disciple of the Enlightenment, uniquely located at the center of great events."[3] A leader in the early days of the French Revolution, he was forced into hiding as the political winds shifted.[4] Concealed in the house of one Madame Vernet in 1793–94, he wrote his most influential work, a little book called *Outlines of an Historical View of the Progress of the Human Mind*. He was eventually arrested and thrown in prison, where he died the day after his arrest, under circumstances that "have been the subject of much speculation ever since."[5] But his final great work was published the next year, and it became a landmark of Enlightenment thinking and would shape how generations understood the idea of progress.

In the book, Condorcet is convinced that the progress of reason has gone too far to allow any future lapse into barbarism, but he is still chagrined at how little progress has been made to increase human happiness. "The friend of humanity," he writes, "cannot receive unmixed pleasure but by abandoning himself to the endearing hope of the future."[6] And what a future Condorcet expects it to be:

> May it not be expected that the human race will be meliorated by new discoveries in the sciences and arts, and, as an unavoidable consequence, in the means of individual and general

> prosperity; by farther progress in the principles of conduct, and in moral practice; and lastly by the real improvement of our faculties, moral, intellectual and physical, which may be the result either of improvement of the instruments which increase the power and direct the exercise of these faculties, or of the improvement of our natural organization itself?[7]

How is this "improvement of our faculties" and "natural organization" to take place? Growing liberty, equality, and prosperity within nations and among nations, Condorcet writes, will raise the general level of instruction, and that in itself will improve human ability. More instruction will in turn produce more knowledge, and Condorcet expects that as we come to understand the world better and improve our ability to teach that understanding, what once might have required genius to uncover or comprehend can become a subject of general knowledge. The result is that we advance the starting point for yet further attainments in the arts and sciences, which in turn increases our powers of action—an upward spiral of enlightenment. As a result, better food, better sanitation, and better medicine will extend the human lifespan, as will "the destruction of the two most active causes of deterioration, penury and wretchedness on the one hand, and enormous wealth on the other."[8] Likewise, Condorcet foresees that "contagious disorders" will be brought under control, as well as occupational and environmental illness. The net result:

> Would it even be absurd to suppose this quality of melioration in the human species as susceptible of an indefinite advancement; to suppose that a period must one day arrive when death will be nothing more than the effect of extraordinary accidents, or of the slow and gradual decay of the vital powers; and that the duration of the middle space, of the interval between the birth of man and this decay, will itself have no assignable limit? Certainly man will not become immortal; but may not the distance between the moment in which he draws his first breath,

> and the common term when, in the course of nature, without malady or accident, he finds it impossible any longer to exist, be necessarily protracted?[9]

Despite his denial of the possibility of immortality, Condorcet expects "that the mean duration of life will for ever increase."[10] It is one of the "laws of nature" that living things are subject to "perfectibility or deterioration."[11] The breeding of animals already shows that we could improve our own physical capacities and senses, but perhaps our "quality of melioration" suggests that a good deal more is possible for us by other means as well.[12] Hence Condorcet wonders whether better-educated parents might not transmit the superior "organization" they have achieved directly to their children.[13]

Condorcet's vision of the future, so familiar to those of us in the West who live it on a daily basis, is very much a reflection of the project laid out more than a century earlier by the English philosopher Francis Bacon for "the enlarging of the bounds of Human Empire, to the effecting of all things possible."[14] Condorcet sees how the increase of our power over nature will soften the hard edges of the human condition by improving the material conditions of life, which will allow at the same time an improvement in the moral conditions. He adds an intimation of accelerating progress; one generation can build on the work of another. A smarter and healthier generation sets the stage for even greater achievement by the next generation. He may have hedged on the question of immortality, but as we shall see, not everyone who came after him was so modest.

Condorcet's thesis—that mankind would improve and expand, growing wiser and healthier and much longer-lived—soon received a powerful rebuttal. Thomas Robert Malthus (1766–1834) was a political economist whose *Essay on the Principle of Population* explicitly attacked Condorcet's depiction of progress.[15] Malthus believed that finite resources limit what human beings can ever hope to accomplish, and that because human reproduction always races ahead of available food, our future holds great misery and scarcity. Malthus's ideas influenced many fields, including biology, where Charles Darwin (1809–1882) adapted them to help explain the workings of evolution, a kind of natural progress caused by competition for limited resources.[16] Through to our own day, much of the debate about progress has arisen from tensions among these three men's ideas: Condorcet's optimism about human perfectibility, the Malthusian problem of resource scarcity, and the Darwinian conception of natural competition as a force for change over time. The transhumanists, as we shall see, reconcile and assimilate these ideas by advocating the end of humanity.

THE INVENTION OF IMMORTALITY

In 1872, the British author and adventurer William Winwood Reade (1838–1875) revisited the project of progress that Condorcet had laid out. Reade, born in Scotland, was a failure as a novelist but had modest success as an African explorer and war correspondent. He was in correspondence with Darwin, who is said to have used in *The Descent of Man* (1871) some information from Reade's expedition to West Africa.[17] While it seems like in his short life Reade never quite lived up to his own expectations for himself, his attempt at a universal history—an 1872 book called *The Martyrdom of Man*—was once highly regarded. W. E. B. Du Bois, Cecil Rhodes, H. G. Wells, and George Orwell all found reasons to praise it.[18] Perhaps even the character Sherlock Holmes was a fan: in *The Sign of the Four*, Holmes says to Watson, "Let me recommend this book—one of the most remarkable ever penned. It is Winwood Reade's *Martyrdom of Man*. I shall be back in an hour."[19] From its original publication to 1910 the book went through eighteen editions in England and seventeen in the United States; one can only imagine that many a Baker Street Irregular has felt compelled to track it down.

The Martyrdom of Man began, Reade says, as an effort to give the hitherto neglected story of "Inner Africa" its due place within European history.[20] But in the writing, the book became very much more: an effort to place human history within a larger natural history that eventually takes Reade right back to the development of the solar system and the origins of life.[21] He adopts his own version of a Darwinian perspective, along with the theories about a geologically dynamic Earth, which were still rather recent in his day. Reade's naturalism is particularly deployed in extensive efforts to provide non-supernatural explanations for the rise of religions.

But there is one crucial aspect of his argument that distinguishes it from most similar presentations in our own day: Reade believes that nature is purposive—and indeed, that something like a cunning of nature is evident in human history.[22] That is to say, human activities like war and religion, or conditions like inequality, serve developmental purposes within a natural scheme of things beyond what is intended by the human beings participating in those activities.[23] "Thus when Nature selects a people to endow them with glory and with wealth her first proceeding is to massacre their bodies, her second, to debauch their minds. She begins with famine, pestilence, and war; next, force and rapacity above; chains and slavery below. She uses evil as the raw material of good; though her aim is always noble, her earliest means are base and cruel. But, as soon as a certain point is reached, she washes her black and bloody hands, and uses agents of a higher kind."[24]

To put it another way, Reade believes that there is a natural imperative for higher abilities and capacities to grow out of lower ones: "The philosophic spirit of inquiry may be traced to brute curiosity, and that to the habit of examining all things in search of food. Artistic genius is an expansion of monkey imitativeness. Loyalty and piety, the reverential virtues, are developed from filial love. Benevolence and magnanimity, the generous virtues, from parental love. The sense of decorum proceeds from the sense of cleanliness; and that from the instinct of sexual display."[25] Reade's claim that the higher derives from the lower does not just apply to human beings. It is a characteristic of life itself, indeed a characteristic of matter, which he regards as inseparable from mind.[26]

We ought not to think there is anything degrading about thus understanding the higher in light of the lower, Reade argues. Indeed, his reductionism opens the door to remarkable possibilities:

> It is Nature's method to take something which is in itself paltry, repulsive, and grotesque, and thence to construct a masterpiece by means of general and gradual laws; those laws themselves being often vile and cruel. This method is applied not only to single individuals, but also to the whole animated world; not only to physical but also to mental forms. And when it is fully realised and understood that the genius of man has been developed along a line of unbroken descent from the simple tendencies which inhabited the primeval cell, and that in its later stages this development has been assisted by the efforts of man himself, what a glorious futurity will open to the human race! It may well be that our minds have not done growing, and that we may rise as high above our present state as that is removed from the condition of the insect and the worm.[27]

That we can assist in our own uplift and greatly transcend what we are today is crucial to Reade's picture of the future, as it is to today's transhumanists. In the natural order of things, the individual human life has limited potential, precisely because by nature we are parts of a whole with at least potentially greater significance:

> As the atoms are to the human unit, so the human units are to the human whole. . . . Nature does not recognise their individual existence. But each atom is conscious of its life; each atom can improve itself in beauty and in strength; each atom can therefore, in an infinitesimal degree, assist the development of the Human Mind. If we take the life of a single atom, that is to say of a single man, or if we look only at a single group, all appears to be cruelty and confusion; but when we survey mankind as One, we find it becoming more and more noble, more and more divine, slowly ripening towards perfection.[28]

That Reade believes mankind is "slowly ripening towards perfection" implies that he is tolerably certain he understands the immediate project that faces humanity, and at least some of its longer-term consequences. Although he claims that he does not mean to suggest that humanity will ever understand the ultimate purpose of creation,[29] he feels confident enough to assert that man was

> not sent upon the earth to prepare himself for existence in another world; he was sent upon earth that he might beautify it as a dwelling, and subdue it to his use; that he might exalt his intellectual and moral powers until he had attained perfection, and had raised himself to that ideal which he now expresses by the name of God, but which, however sublime it may appear to our weak and imperfect minds, is far below the splendour and majesty of that Power by whom the universe was made.[30]

By the power of science rather than prayer, Earth, "which is now a purgatory, will be made a paradise."[31] The genuinely "Sacred Cause" is "the extinction of disease, the extinction of sin, the perfection of genius, the perfection of love, the invention of immortality, the exploration of the infinite, the conquest of creation."[32]

So by making men mortal and immoral, nature points humanity in the direction of immortality and morality so long as we exercise our intelligence.[33] Reade could already see signs of progress in this direction: "Life is full of hope and consolation; we observe that crime is on the decrease, and that men are becoming more humane. The virtues as well as the vices are inherited; in every succeeding generation the old ferocious impulses of our race will become fainter and fainter, and at length they will finally die away."[34] Delusions about an immortal soul will only stand in the way of such efforts; Christianity, which Reade treats under general headings such as "Religion" and "superstition," will have eventually done the work intended of it as a tool of nature, and at that point can and must be destroyed, for it is in the nature of these tools to become obstructions once they have brought life to the next level.[35] While human beings may never rival the great Creator of all things, there is a long way to go before that would become an issue.[36] Echoing Francis Bacon,[37] Reade notes that "we can conquer Nature only by obeying her laws, and in order to obey her laws we must first learn what they are. When we have ascertained, by means of Science, the method of Nature's operations, we shall be able to take her place and to perform them for ourselves."[38] Nature intends that we rebel against being the serfs of nature.[39]

Having placed immortality explicitly on the agenda of the future, Reade considers space travel a necessary consequence:

> Disease will be extirpated; the causes of decay will be removed; immortality will be invented. And then, the earth being small, mankind will migrate into space, and will cross the airless Saharas which separate planet from planet, and sun from sun. The earth will become a Holy Land which will be visited by pilgrims from all the quarters of the universe.[40]

When human beings "invent" immortality we press beyond the natural order of things in which we are mere cells in a larger whole, and when we die are dead forever. In similar fashion, with space travel we will also have proven the essentially mundane character of the once-transmundane heavens. Yet Reade's imaginative assurance about the great things ahead for humanity puts in high relief the ignorance and miseries of humanity today.

> These bodies which now we wear belong to the lower animals; our minds have already outgrown them; already we look upon them with contempt. A time will come when Science will transform them by means which we cannot conjecture, and which, even if explained to us, we could not now understand, just as the savage cannot understand electricity, magnetism, steam.[41]

This is a glorious future, one in which men will be "perfect," having the power of "what the vulgar worship as a god."[42] But Reade recognizes—more so than did Condorcet—that this vision is not entirely consoling. In a prayer-like passage, he acknowledges that it makes the ills of the present look all the more terrible, and the past a yet darker place:

> You blessed ones who shall inherit that future age of which we can only dream; you pure and radiant beings who shall succeed us on the earth; when you turn back your eyes on us poor savages, grubbing in the ground for our daily bread, eating flesh and blood, dwelling in vile bodies which degrade us every day to a level with the beasts, tortured by pains, and by animal propensities, buried in gloomy superstitions, ignorant of Nature which yet holds us in her bonds; when you read of us in books, when you think of what we are, and compare us with yourselves, remember that it is to us you owe the foundation of your happiness and grandeur, to us who now in our libraries and laboratories and star-towers and dissecting-rooms and work-shops are preparing the materials of the human growth. And as for ourselves, if we are sometimes inclined to regret that our lot is cast in these unhappy days, let us remember how much more fortunate we are than those who lived before us a few centuries ago.[43]

The fact that he calls his book *The Martyrdom of Man* indicates that Reade is well aware of the tragic side of his progressivism. But "in each generation the human race has been tortured that their children might profit by their woes. Our own prosperity is founded on the agonies of the past. Is it therefore unjust that we also should suffer for the benefit of those who are to come?"[44] Until men become immortal, the only satisfaction to be found is in the superiority of the present to the past, and the chance of making one's own infinitesimal contribution to the future.

SALVATION IN SPACE

We turn next to another thinker who concluded early on that humanity, in order to preserve itself, would have to venture into outer space. Nikolai Fedorovich Fedorov (1829–1903), the illegitimate son of a Russian prince, was an intense but retiring Moscow librarian who was "reputed to have read all the books he catalogued."[45] Unlike all the other figures discussed in this chapter, Fedorov was not widely known during his own lifetime. The posthumous publication of two volumes of his work did not change that situation a great deal, even though the publisher made them available free of charge, in accord with Fedorov's beliefs about property.[46] But the quality of those who admired his work makes up for the lack of quantity: he was known and respected by both Dostoevsky and Tolstoy. And Fedorov had one yet more important connection: he assisted, and some think passed his ideas on to, Konstantin Tsiolkovsky, among the greatest of the pioneers of space travel.[47]

In a work composed sometime after a famine in 1891, Fedorov writes that the "learned" have neglected their obligation to the "unlearned" to improve the conditions of their lives, particularly the lives of agricultural workers.[48] He is mightily impressed by reports of using explosives to create rain. He regards using the tools of war for peaceful purposes as literally providential, a sign of what God expects of man.[49] He is much more skeptical than Reade about the cunning of nature, asserting that it is "extreme childishness" to expect that the "blind force" of nature will produce just good results. It is only when human beings put their will behind their common task—to understand and control that force—that it will be turned to the good by our conscious control.[50]

Fedorov is well aware that such control as we currently possess is far from guaranteed to be used for the benefit of mankind. What is lacking, he believes, is the necessary sense of human kinship.[51] "Unbrotherly relations" make life the "struggle" that has hitherto been definitive of human civilization.[52] They also lead to a thoroughgoing misunderstanding of the true meaning of progress. Progress, Fedorov claims, is not to be seen in the superiority of man over beast, the superiority of the present generation over past generations (as Reade would say), or the superiority in this generation of the young over the old.[53] Indeed, such a picture of progress has its tragic tone because it is inherently divisive: "Progress makes fathers and ancestors into the accused and the sons and descendants into judges; historians are judges over the deceased, that is, those who have already endured capital punishment (the death penalty), while the sons sit in judgment over those who have not yet died."[54]

In contrast, Fedorov—a devout if unconventional adherent of Russian Orthodoxy—writes that we should take our cue from "true religion," which is "the cult of ancestors, the cult of all the fathers as one father inseparable from the Triune God, yet not merged with him."[55] On this basis Fedorov imagines the single, common task of mankind as a union of sons bent on overcoming the blind forces of nature—not only to defeat hunger, disease, and death for the living, but to achieve the resurrection of all of their fathers, grandfathers, great-grandfathers, etc.[56] In a passage that could have been written by a number of today's transhumanists, he says, "Death has become a general organic evil, a monstrosity, which we no longer notice and no longer regard as an evil and a monstrosity."[57] But we will learn to bring the dead back to life, "substituting resurrection for birth"; we will thereby eliminate the need for sexual reproduction, which is just another example of the blind operation of nature.[58] We will solve hunger, too, substituting "creativity for nutrition": we will not need to eat, but will produce ourselves "out of the very basic elements into which the human body can be decomposed."[59]

Even so, Fedorov is not unconcerned about the Malthusian problem of eventual exhaustion of resources here on Earth. He has his own, to us familiar, apocalyptic vision:

> The extinction of stars (sudden or slow) is an instructive example, a terrifying warning. The growing exhaustion of the soil, the

> destruction of forests, distortions of the meteorological process manifested in floods and droughts—all this forebodes 'famines and plagues' and prompts us to heed the warning. Apart from a slowly advancing end, we cannot be certain whether a sudden catastrophe may not befall the Earth, this tiny grain of sand in the vastness of the Universe.[60]

For such reasons resurrection will not suffice; the exploration of outer space is also absolutely necessary to prepare the "future homes of the ancestors."[61] God has arranged that "the Earth itself has become conscious of its fate through man" and this consciousness would be useless were we simply to stand by and observe "the slow destruction of our home and graveyard" at the hands of purposeless nature.[62] Rather, "God is the king who does everything for man but also through man" and he intends that humanity not be "idle passengers" but "the crew of its terrestrial craft"[63]—a remarkable prefiguration of the space-age/environmentalist idea of "Spaceship Earth."

The reasons for developing space travel transcend the merely practical necessities of overcoming resource exhaustion ("the economic problem posed by Malthus"[64]) and finding a place for resurrected ancestors. Space travel will also deeply affect our moral nature. Is it more fantastic, Fedorov asks, to believe in the Christian understanding of heaven and the afterlife—"to create a moral society by postulating the existence of other beings in other worlds and envisioning the emigration thither of souls, the existence of which cannot be proven"—or to believe that we might visit other worlds ourselves someday?[65] Moreover, to "inhabit all heavenly bodies" would

> unite all the worlds of the Universe into an artistic whole, a work of art, the innumerable artists of which, in the image of the Triune Creator, will be the entire human race . . . attaining divine perfection in the cause, the work of restoring the world to the sublime incorruptibility it had before the fall. Then, united, science and art will become ethics and aesthetics; they

> will become a natural universal technology of their work of art.[66]

It is through human efforts, then, that a *cosmos*—an ordered universe—comes to exist, rather than the purposeless *chaos*, the mere raw materials given us by God, that exists prior to human intervention. We will see a similar point being made many times in the pages that follow.

Both Fedorov and Reade seem to assume that outer space is entirely available for human colonization. In so doing they are taking a position on the question of the existence of life on other worlds, which was in fact already much debated in their day. Developments in astronomy, geology, and biology, in particular the thinking of Darwin, were leading many to consider that there was no reason to expect Earth to be the sole abode of life—and if life could exist elsewhere, so too intelligence may have evolved. We turn to one such thinker next.

EDIFYING ALIENS

French astronomer and spiritualist Nicolas Camille Flammarion (1842–1925) ranks high among the great popularizers of science, particularly astronomy—the Carl Sagan of his time. Owner of a private observatory and author of some seventy books, he was particularly interested in the possibility of extraterrestrial life, and was among the first to imagine it in thoroughly alien terms, an idea he presented in both his popular science writing and science fiction. A sense of his contribution can be gauged by the fact that he has craters named for him on both the moon and Mars. Into the 1960s, decades after his death, his name remained on a popular, if by that time much updated, introduction to astronomy. And he garnered all this recognition despite the fact (perhaps in his own time because of the fact?) that he considered spiritualism to rank among his scientific interests.

Writing early in the twentieth century, Flammarion articulates what is often called the "assumption of mediocrity": that there is nothing special about Earth's place in the universe, and so life likely exists elsewhere. "On what pretext could one suppose that our little globe which, as we have seen, has received no privileges from Nature, is the exception; and that the entire Universe, save for one insignificant isle, is devoted to vacancy, solitude, and death?"[67] If outer space looks unfriendly to life at first glance, we learn even from terrestrial observation that the environment that is hostile to one form of life is favorable to another. So the fact that extraterrestrial conditions may appear hostile to life as we know it is no warrant that life is impossible.

Furthermore, Flammarion argues that extraterrestrial life will likely be different from life as we know it, perhaps with a different chemical basis and very different capabilities. Perhaps it will in some respects be superior to us. Might not nature

> have given to certain beings an electrical sense, a magnetic sense, a sense of orientation, an organ able to perceive the ethereal vibrations of the infra-red or ultra-violet, or permitted them to hear at a distance, or to see through walls? We eat and digest like coarse animals, we are slaves to our digestive tube: may there not be worlds in which a nutritive atmosphere enables its fortunate inhabitants to dispense with this absurd process? The least sparrow, even the dusky bat, has an advantage over us in that it can fly through the air. Think how inferior are our conditions, since the man of greatest genius, the most exquisite woman, are nailed to the soil like any vulgar caterpillar before its metamorphosis! Would it be a disadvantage to inhabit a world in which we might fly whither we would; a world of scented luxury, full of animated flowers; a world where the winds would be incapable of exciting a tempest, where several suns of different colors—the diamond glowing with the ruby, or the emerald with the sapphire—would burn night and day (azure nights and scarlet days) in the glory of an eternal spring; with multi-colored moons sleeping in the mirror of the waters, phosphorescent mountains, aerial inhabitants—men, women, or perhaps of other sexes—perfect in

> their forms, gifted with multiple sensibilities, luminous at will, incombustible as asbestos, perhaps immortal, unless they commit suicide out of curiosity?[68]

For Flammarion, the possibility of life elsewhere—indeed, life everywhere—raises the question of the ultimate destiny of human life. For it is life, he believes, not matter, that is the key to understanding the universe, yet life eventually seems to give way to mere matter; individually we are built on death and proceed unto death ourselves, and the same is true for our planet as a whole. Hence "Let no one talk of the Progress of Humanity as an end! That would be too gross a decoy."[69] Flammarion believes that progress is the law of life,[70] but, as Fedorov also suggested, *material* progress alone would mean that in the end we would still fall prey to entropy itself—that each of our lives, and that human life as a whole, will be extinguished. We reject this gloomy possibility, he says, as being "incompatible with the sublime grandeur of the spectacle of the universe."[71] So while "Creation does not *seem* to concern itself with us,"[72] this appearance may be deceiving. He even goes so far as to ask if "distant and unknown Humanities"—that is, alien races—might not be "attached to us by mysterious cords, if our life, which will assuredly be extinguished at some definite moment here below, will not be prolonged into the regions of Eternity."[73]

One would be hard pressed today to find a popularizer of science who, like Flammarion in this poetic and confusing passage, seems to hover between a fairly traditional notion of heaven and a suggestion of interstellar reincarnation. But he seems to have felt that our intuition that our lives cannot end had support from his astronomy: "As our planet is only a province of the Infinite Heavens, so our actual existence is only a stage in Eternal Life. Astronomy, by giving us wings, conducts us to the sanctuary of truth. The specter of death has departed from our Heaven. The beams of every star shed a ray of hope into our hearts."[74]

For Reade and Fedorov, the prospect of a universe that can be enlivened by human action, which involves transcending the natural order even to the point of inventing immortality, gives hope and meaning to the human future. For Flammarion, on the other hand, the conclusion that alien life is already omnipresent and diverse calls attention to the parochialism of our own view of ourselves, fostering scientific imagination of beings different from and even superior to us. We see the limitations on our own lives by imagining beings with different capacities, and in that light our own limits appear simply arbitrary. Yet at the same time, the likelihood of alien life suggests to him the insignificance of brute matter. The ascendance of life that he imagines we will see in the universe as a whole opens the door to a hope for the yet greater victory for life that would be personal immortality.

BECOMING ALIENS

Our next thinker shared Fedorov's and Flammarion's concern about the limited prospects for material progress if we are confined to Earth, and so believed that humanity was destined to explore outer space—but that, in the long term, we would become something no longer recognizably human at all. J. B. S. Haldane (1892–1964) was a distinguished scientist, a major public intellectual, and an outspoken Marxist. His main contributions to science were in quantitative analysis of genetics and evolutionary biology—from which one would not necessarily adduce the humor, imagination, and charm of his popular writing. His influence was such that he helped inspire two of the greatest works of anti-utopian literature of the twentieth century: his acquaintance Aldous Huxley based some of *Brave New World* on Haldane's ideas, and C. S. Lewis is said to have had Haldane in mind for various speeches and characters in his space trilogy.[75]

Haldane most vividly sketched out his vision of the human future in "The Last Judgment," a piece of writing from 1927 that is part essay, part science fiction story.[76] (Lewis considered it "brilliant, though to my mind depraved."[77]) Haldane begins by looking at how the Earth and our sun might come to their natural ends. He soon turns to consider a theme that we will see become increasingly common, how we might destroy the Earth ourselves, imagining an account of the last millennia of human life on Earth as it might be told by a distant descendant living on Venus.[78] The premise of this story is that as a consequence of having "ridiculously squandered" tidal power over a period of some five million years, humans have changed the moon's orbit until it comes so close to Earth that it is pulled apart, in the process making Earth uninhabitable.[79]

While this result was long predictable, humans "never looked more than a million years ahead" so few were ever concerned with this consequence of using tidal power.[80] Instead, in the course of their three-thousand-year-long lives, most people concentrated on "the development of personal relationships" and on "art and music, that is to say, the production of objects, sounds, and patterns of events gratifying to the individual."[81] Natural selection having ceased, the only substantial change to humanity was "the almost complete abolition of the pain sense."[82] Real advances in science came to a halt; rather than try to develop the human race, attention was paid to breeding beautiful flowers.

Having foreseen what was to come, however, a few did what they could to assure the existence of life elsewhere after the anticipated disaster. That is no small task as Haldane paints it; even simple steps like managing to land explorers successfully on the moon, Mars, and Venus takes a couple of million years. The technical difficulties of landing and return are compounded by the disinclination of individualistic humans to give up their long lives on what amount to suicide missions. Those who finally land alive on Mars are destroyed by sentient alien life already established there, as Flammarion might have expected. Those who land on Venus find extremely hostile environmental conditions under which humans cannot possibly survive. Efforts at further exploration are dropped for the time being.

About eight million years later, the approaching moon having disrupted earth's geology and ecosystems, a minority undertake renewed efforts to colonize Venus. "A few hundred thousand of the human race . . . determined that though men died, man should live forever."[83]

By a ten-thousand-year-long effort at selective breeding, humans create a new race that can survive on Venus.[84] These colonists are sent out in 1,734 ships; eleven manage to land.[85] Such life as Venus already had, inimical to the colonists, is utterly destroyed by bacteria prepared for that purpose. From that point on, the settlement of Venus proceeds apace.

Our Venusian descendants were designed by the small minority of species-minded Earthlings to share a hive mind; they do not suffer from the selfish propensity for seeking individual happiness that led to Earth's destruction. Two new senses contribute to the hive mind: at every moment they sense "the voice of the community," and they also have a sort of built-in radio that can be turned on or off at will. They are also genetically predisposed to look to the future more than the past, unlike Earthlings whose strange backward-looking propensities are illustrated not only by their failure to act in the face of their destructive tendencies, but by their religious beliefs. The Venusians' forward-looking characteristic also makes them more willing to sacrifice themselves.

The net result is that the Venusians see their potential extending far beyond anything humans ever could have accomplished; "we have settled down as members of a super-organism with no limits on its possible progress."[86] They plan to breed a version of themselves that will be able to settle Jupiter. Foreseeing in 250 million years an improved opportunity for interstellar travel, they think they can take it "if by that time the entire matter of the planets of the solar system is under conscious control."[87] Only a few of the millions of projectiles they send out might succeed. The Venusians are undaunted:

> But in such a case waste of life is as inevitable as in the seeding of a plant or the discharge of spermatozoa or pollen. Moreover, it is possible that under the conditions of life in the outer planets the human brain may alter in such a way as to open up possibilities inconceivable to our own minds. Our galaxy has a probable life of at least eighty million million years. Before

> that time has elapsed it is our ideal that all the matter in it available for life should be within the power of the heirs of the species whose original home has just been destroyed. If that ideal is even approximately fulfilled, the end of the world which we have just witnessed was an episode of entirely negligible importance. And there are other galaxies.[88]

In his commentary on his story at the end of "The Last Judgment," Haldane acknowledges that he is not really trying to predict the future—he is just engaging in an imaginative thought experiment, a "valuable spiritual exercise."[89] The future will certainly not conform to our present ideals, but thinking about it can illuminate "our emotional attitude towards the universe as a whole" that presumably is one source of those ideals.[90] Traditionally, that attitude has been the province of religion. But modern science has taught us that the universe is far vaster in size and possibilities than religions ever knew, and so it is necessary to start using our imaginations in connection with these new realities. In effect, then, science fiction stands in for religion. The new scale of things we can begin to imagine should call forth a greater ambition among the most creative humans to develop (and for the rest of us to cooperate in) a plan that goes beyond traditional ideas of salvation, such as the assumption that the purpose of creation is to prepare some few for "so much perfection and happiness as is possible for them."[91]

We can "only dimly conjecture" what this plan might be, but Haldane wonders whether it might be the "emergence of a new kind of being which will bear the same relation to mind as do mind to life and life to matter."[92] As we can already envision the end of our own world, some such transformation will be necessary. Only if the human race proves that "its destiny is eternity and infinity, and that the value of the individual is negligible in comparison with that destiny," will "man and all his works" not "perish eternally."[93]

The tension within this edifying conclusion is not hard to spot. In Haldane's scheme, an eternal and infinite destiny can only be achieved by making man himself into one of the works of man, such that in fact *human* beings *do* perish eternally. Furthermore, the imagination of this superior progeny is really an exercise in elucidating all the reasons for which, by and large, we should not be missed. So Haldane's substitute for religion embodies an "emotional attitude towards the universe as a whole" which is predicated on the assumption that whatever human beings do, "man's little world will end."[94] The real choice is between ending it ourselves and having it ended for us, or perhaps between ending it accidentally and ending it deliberately.

MANKIND REMANUFACTURED

Haldane may have claimed he was not trying to predict the future, but our final author certainly was—and in the process, he lays out a rather specific path toward what he calls "the progress of dehumanization," integrating many of the themes that our other authors developed.

The Irish-born J. D. Bernal (1901–1971) was, like Haldane and Flammarion, a scientist by training. He is probably best known for the development of the mathematics of X-ray crystallography, which quickly became a key technique of chemical analysis.[95] (It was this technique that allowed the double-helix structure of DNA to be discovered, for example.) He did research that helped facilitate the D-Day landings, and made serious contributions to the sociology of science.[96] A public intellectual of some note, Bernal was a dedicated communist and admirer of the Soviet Union; in 1953, he was awarded the Stalin Peace Prize, a prominent Soviet prize for the country's international supporters, and from 1959 to 1965 he was president of the World Peace Council, a Soviet-funded international activist group.[97]

Bernal's first popular publication was a thin volume called *The World, the Flesh and the Devil: An Inquiry into the Three Enemies of the Rational Soul* (1929).[98] In it, he proposes an objective effort to predict things to come. Bernal acknowledges that this task might be easier said than done, partly because it can be difficult to distinguish prediction and desire, partly because of all the complex interactions that make the world what it is, and partly because "all evidence" points "to ever increasing acceleration of change."[99] Nevertheless, it is reasonable to start by looking at what the trends are. Bernal projects the future in three areas: "the world," or our power in relationship to the material world; "the flesh," or our power over life, particularly our own bodies; and "the devil," our power over our own psyches. He concludes his volume by attempting to see what might come of developments in these three areas taken together.

In each realm Bernal expects remarkable things. His chapter "The World" focuses primarily on "the conquest of space."[100] He anticipates some developments that we have only recently achieved, like the use of huge sails to propel ships using solar wind,[101] but his most extended discussion is of what it would take to create ten-mile-diameter spherical habitations with tens of thousands of inhabitants.[102] With the necessary propulsion systems added, these communities in space would eventually allow for the long voyages that interstellar travel would require—voyages that will be necessary as our sun begins to fail.[103]

The chapter "The Flesh" starts from the bald assertion that "modern mechanical and modern chemical discoveries have rendered both the skeletal and metabolic functions of the body to a large extent useless."[104] Bernal expects the increasing substitution of mechanical for biological systems in the human body, with all the augmentation of physical and sensory abilities that implies—for example, "we badly need a small sense organ for detecting wireless frequencies."[105] People have always wanted longer lives and more opportunities "to learn and understand."[106] But achieving such goals is now in sight:

> Sooner or later some eminent physiologist will have his neck broken in a super-civilized accident or find his body cells worn beyond capacity for repair. He will then be forced to decide whether to abandon his body or his life. After all it is brain that counts, and to have a brain suffused by fresh and correctly prescribed blood is to be alive—to think. The experiment is not

> impossible; it has already been done on a dog and that is three-quarters of the way towards achieving it with a human subject.[107]

Bernal expects that once some men were thus transformed, they would be most able at transforming others.[108] Humans will have a "larval" stage of six to twelve decades in our current bodies, then we will pass into "chrysalis, a complicated and rather unpleasant process of transforming the already existing organs and grafting on all the new sensory and motor mechanisms."[109] Of course, unlike a butterfly, the end result of the human transformation will be capable of constant upgrade and modification—and indeed there will be no one form into which people will change themselves in any case, as the mechanical body will be readily customizable.[110]

"Normal man is an evolutionary dead end; mechanical man, apparently a break in organic evolution, is actually more in the true tradition of a further evolution."[111] Bernal envisions each of these mechanical men as looking something like a crustacean, with the brain protected in a rigid framework and a system of appendages and antennae attached for sensing and manipulating the world.[112] He freely acknowledges that, to us, these beings would appear "strange, monstrous and inhuman."[113] But he claims that such monsters are "only the logical outcome of the type of humanity that exists at present."[114]

In any case, beings so designed would quickly become progressively more different from us. Their brains would be readily linked together electronically to become a kind of group mind. Thus, while the original individual organic brain itself would still have a limited lifespan (perhaps three hundred to a thousand years, Bernal estimates), sharing its feelings, knowledge, and experience with other brains would be a way of "cheating death."[115] Bernal ends the chapter with the speculation that these inhuman beings would invent whole new materials and forms of life out of which to constitute themselves, so that even organic brain cells could be replaced with more diffuse materials with more complex interconnections, thus ensuring itself "a practical eternity of existence."[116] They might transcend physical embodiment altogether, becoming completely etherealized, atoms in space communicating by radiation, ultimately perhaps resolving entirely into light. "That may be an end or a beginning, but from here it is out of sight."[117]

These first chapters lay out what would become an agenda for decades of science fiction and a fair amount of actual research and development. The next chapter, "The Devil," is one that Bernal himself expressed dissatisfaction with nearly four decades after the book was published; he admitted that it was too much written under the influence of Freud.[118] But the issue it discusses remains one that is debated, even if not precisely on Bernal's terms. The main question is whether continued progress in science will be able to overcome the problem posed by the new (that is, Freudian) insight that "the intellectual life" is not "the vocation of the rational mind, but . . . a compensation . . . a perversion of more primitive, unsatisfied desires."[119] That is, science requires an ongoing supply of "perverted individuals capable of more than average performance."[120] Should our psychology and our power over nature combine to make the satisfaction of our desires the norm, we could settle into a "Melanesian" life of "eating, drinking, friendliness, love-making, dancing and singing, and the golden age may settle permanently on the world" without any desire for further progress at all.[121] (Note the similarity here with Haldane's flower breeders.) On the other hand, it could also be that we might be able to live lives that are both "more fully human and fully intellectual" if "a full adult sexuality would be balanced with objective activity."[122]

The question of whether all this progress will eliminate the desire for further progress has another side as well, given the "distaste" that Bernal acknowledges he feels, and others are likely to feel, about what the future holds "especially in relation to the bodily changes."[123] It may even be that people will not have a chance to get used to such changes gradually, given the accelerating rate of change. Bernal does not pretend to predict whether repugnance, combined with satisfaction, will ultimately triumph over the increasing power that will be in the hands of those who advance the cause of science and mechanization. But one result might be the "splitting of the human race" into two branches: a stagnant because "fully balanced humanity" and another branch "groping unsteadily beyond it."[124] Seeing how that outcome might arise is the point of the book's concluding chapters.

Bernal's basic thought here is that the mechanical men he envisions would be very well suited for colonizing and exploiting space, as their life-support requirements would be far less than what human beings require and their capacities would be wider.[125] He imagines these transhumans as "connected together by a complex of ethereal intercommunication" and spread out across space and time.[126] But he is brought up short by the recognition that the human mind had hitherto "evolved always in the company of the human body."[127] The radical change he anticipates to "the delicate balance between physiological and psychological factors" will create "dangerous turning points and pitfalls."[128] What will happen to the sexual drive, for example? Perhaps it will require yet more thoroughgoing sublimation into research or, even more likely, into "aesthetic creation."[129] As these new beings come ever more completely to understand the world around them, and ever more capable of manipulating it, their primary purpose is likely to become determining "the desirable form of the humanly-controlled universe which is nothing more nor less than art."[130]

After much consideration about how the possibility of "permanent plenty" might transform society, Bernal settles on the thought that the future is likely to hold *de facto* or even *de jure* rule by a scientific elite that could be the first stable aristocracy.[131] This elite would have the means to assure that the masses engage in "harmless occupations" in a state of "perfect docility under the appearance of perfect freedom."[132] "A happy prosperous humanity enjoying their bodies, exercising the arts, patronizing the religions, may be well content to leave the machine, by which their desires are satisfied, in other and more efficient hands."[133]

Since Bernal thinks that those who tend the machines will increasingly be machines themselves, we now see why he thinks the human race might split into two branches. Yet whereas it seems very likely, as he has suggested, that the distinction between machine-men and men would also be the distinction between ruler and ruled, perhaps that would not have to be the case. For as Bernal notes, science depends on the supportive routine work of non-scientists, and on the recruitment from the many of the few most capable minds. Furthermore, he claims, scientists themselves tend to have a strong identification with humanity. So the first stable aristocracy could be a meritocracy that might at least recruit (or should one say harvest?) fresh brains from the most promising of humans. Still, characteristics that might bind the two groups will likely diminish with time, allowing the underlying processes producing dimorphism to hold sway.[134] At that point it is quite possible that "the old mechanism of extinction will come into play. The better organized beings will be obliged in self-defense to reduce the numbers of the others, until they are no longer seriously inconvenienced by them."[135] The main hope for a different outcome is once again the prospect that the more advanced beings will settle in space, leaving Earth to the old-fashioned model in "a human zoo, a zoo so intelligently managed that its inhabitants are not aware that they are there merely for the purposes of observation and experiment."[136] So decades before today's transhumanists, Bernal predicts the survival of humanity as a curiosity (at best).

Bernal is not certain his vision will prevail,[137] nor does he hold that the developments he lays out will produce a perfect world:

> the dangers to the whole structure of humanity and its successors will not decrease as their wisdom increases, because, knowing more and wanting more they will dare more, and in daring will risk their own destruction. But this daring, this experimentation, is really the essential quality of life.[138]

Bernal's predictions are not so millennial as Fedorov's, nor as overtly tragic as Reade's. But they do contain tensions. He recognizes that his scientifically driven "progress of dehumanization" is motivated by an ultimately unfulfillable desire for the mysterious and supernatural.[139] Those who push the boundaries of knowledge outward will create a world that, for them at least, will be ever more prosaic, and therefore of less interest. Even if this process is infinite, it retains a Sisyphean character, and one might wonder: why bother? It turns out, however, that this "daring" effort to transcend one's time and control one's life is nothing other than an expression of life itself that is beyond our control. There is a natural fatality to our effort to control nature.

FROM BETTER HUMANS TO BEYOND HUMANITY

My purpose in presenting these examples of thinkers who anticipated today's transhumanism is not to suggest how these thinkers might have influenced one another, or to prove their influence on today's transhumanists. But I trust it is reasonably clear that between Condorcet and Bernal the idea of progress itself has traveled quite a distance. Where for Condorcet the friend of humanity can find reason to think that in the world to come people will be more humane to each other, when Bernal looks to the future he sees "the progress of dehumanization": human extinction at worst, and at best human irrelevance to the progressive development of intelligence and power over the natural world—an evolutionary "dead end." As we will discuss later, Oxford philosopher Nick Bostrom, one of the founders of the World Transhumanist Association and the director of Oxford's Future of Humanity Institute, thinks we can have it both ways, expressing the hope that our posthuman replacements will be designed to be *more* humane than we are. Everyone seems to agree that the kinder, gentler world that Condorcet imagines could indeed come into being; but for Haldane it would amount to a short-sighted squandering of nature's potential, while for Bernal we humans would be likely to live in it as subjects to powers so far beyond our control as to make historical aristocracies seem models of egalitarianism. Today, David Pearce, another founder of the World Transhumanist Association, cuts through the problem of needing some to rule others in order to keep them happy by suggesting that we can redesign ourselves so that we are always experiencing "a sublime and all-pervasive happiness."[140] Condorcet's expectation that people will be better fed becomes the revolt in Fedorov and Flammarion against the "absurd" need to eat at all, while today's transhumanists likewise find it unacceptable that we eat and excrete as we do.[141] Condorcet suggests the possibility of accelerating progress toward the conquest of nature here on Earth, but for Haldane and Bernal what is at stake is the aesthetic recreation of the universe itself. Today, inventor and author Ray Kurzweil, the most widely known of the transhumanists, wonders if posthuman superintelligence might not be able to overcome entropy itself, thereby preventing the now-expected eventual end to the possibility of life in the universe and overcoming the last challenge to the immortality that Condorcet was only willing to hint at.

There is no single arc that connects all of our authors in such a way as to account for this significant transformation in the understanding of progress. But each lays part of the foundation for the change, a foundation on which is built the edifice that is contemporary transhumanism. Let us try to identify some of the key points.

The main line of Condorcet's argument is the most familiar. Perhaps building off Rousseau's notion of human "perfectibility,"[142] Condorcet asserts that we possess a (unique?) "quality of melioration" that allows us to improve ourselves, primarily through the conquest of nature. As that project succeeds, many of the longstanding, seemingly given conditions of human life—poverty, hunger, disease, vice, and other pervasive disabilities that have stood in the way of a good human life—become problems that can be solved. By solving them, we make better human beings, human beings who are physically more fit, mentally more capable, morally improved. The rate of progress accelerates with this new starting point, but Condorcet seems to believe that people will remain human beings.

Yet Condorcet's own ideas about life extension, combined with the possibility of accelerating progress, begin to suggest something more radical. Initially life extension seems of a piece with the other improvements he speaks of. After all, as we become healthier and eat better, longer life would seem to follow as a matter of course. One could say that Condorcet is merely pointing out how we can reduce the incidence of premature death. But when he starts talking about an indefinite extension of lifespan, the door is opened to the possibility of a significantly more fundamental change in the terms of human existence. It does not seem as if Condorcet wishes to open the door very wide. Yes, he looks forward to a time when death becomes something that is chosen, but note that it is chosen "in the course of nature," as if there is in this respect at least some part of nature that human beings will not or should not master.[143]

This limit is one that our other authors are not nearly so inclined to respect, and their overt desire for immortality is of a piece with a far stronger inclination to imagine the development out of humanity of some completely new kind of superior being. Perhaps Condorcet did not understand, or chose not to highlight, the more radical consequences of his own picture of accelerating progress. But it seems more likely that something had to be added to Condorcet in order to promote this shift in the imagination. What might that be?

First, as Fedorov explicitly highlighted, there is the Malthusian dilemma of resource scarcity. Malthus wrote in direct response to Condorcet's hopes for the future, arguing that the melioration Condorcet imagined would be self-defeating. More people living materially more comfortable lives will simply produce resource scarcity, which in turn will bring back all the ills of human life. Perpetual progress understood as an ongoing improvement in the material conditions of life for all is thus simply impossible since population will grow faster than available resources. For Fedorov, the conquest of space is in part a solution to the Malthusian dilemma of resource exhaustion, a solution that becomes the more plausible as the sense grows that the Earth is but a tiny speck in a very large universe.[144] Of course, as the hostility of the extraterrestrial environment came to be better understood, this prospect may have seemed more daunting. But Haldane and Bernal are there with a solution to this part of the problem: the radical reconstruction of humanity in a way that makes mankind better suited for life away from Earth.

As we have seen, once you have taken that step, the promise of effectively infinite worlds in infinite space makes anything seem possible; both Haldane and Bernal take us to the very limits of the human imagination, Haldane by suggesting that all the matter in the galaxy available to life should be used by it before moving on to other galaxies, Bernal by suggesting that our distant descendants will remake the universe with a new "let there be light." Perhaps these beings that conquer space will also conquer time and entropy, a route to the eternity promised by Flammarion and the end to death promised by Fedorov.

The resource scarcity that Condorcet did not worry about implies ongoing competition among human beings rather than an ever more cooperative world, and by the late nineteenth century that ongoing competition was firmly associated with Darwinian evolution. This intellectual revolution is the second change that has pushed Condorcet's successors in a more radical direction. Condorcet could assert that living things must be on a course of either perfection or decline; human beings in the future could change, but the result would be perfected or degraded human beings. From a Darwinian point of view, as Reade highlights, why should there not be changes in the future that correspond in magnitude to the changes that produced man as we now observe him? After Darwin, it becomes possible, *if not downright necessary*, to think that future human descendants will not be human.

Scholars disagree about whether Darwin himself conceived that the evolutionary changes that brought about human beings (and other species) should be called "progress" or merely change. There is agreement that some of his writings point in one direction, some in another. At present, the "mainstream" scholarly view is that Darwin's statements implying that there is an ascent to humanity were mere concessions to the progressive spirit of his Victorian times, and that Darwin himself understood that his principles allowed him to speak of evolutionary change but not progress. Yet there is also an impressive body of arguments and evidence to suggest that Darwin *did* believe in evolutionary progress, so long as we take sufficient care to define what that phrase means.[145] In either case, it seems indisputable that a great many of those who, like Reade, were influenced by Darwin's ideas *took* him to be pointing to an evolutionary ascent to humankind. And if so, why should that process not continue to produce yet higher forms of life? After all, even the penultimate words of Darwin's *Descent of Man* suggest that man may be excused for "feeling some pride at having risen, though not through his own exertions, to the very summit of the organic scale," which in turn may give "hopes for a still higher destiny in the distant future."[146]

Yet all of our thinkers would agree that there will be a difference between the blind evolution that produced humanity and the future evolution driven by human beings and then posthumans, precisely because our "own exertions" can now play a part. If evolution is the law of life, and at the same time if evolution has brought about human beings who can take hold of evolution and direct it as one more aspect of our control over nature, then a grand narrative of free human creativity becomes possible. Our conquest of nature is no longer a local affair but takes on a cosmic significance. For Reade, Fedorov, or Haldane's Venusians, the creation of new forms of life, the enlivening of the cosmos, is the goal of goals, the highest good. We transform the universe by transforming ourselves.

Flammarion may be the superior Darwinian here, thinking that it is at least short-sighted and at worst completely inconsistent to think that all this marvelous development of matter into life and life into intelligence should be a process confined to one planet alone. Yet he does not highlight the tougher Darwinian consequences of this line of thought. What if the cosmos is not ours to do whatever we want with, because other forms of life have already staked a claim? Haldane seems to understand the situation best of all our thinkers: he extends the realm of competition beyond our world, with the expected consequences of extraterrestrial winners (the Martians in Haldane's tale) and losers (the aboriginal Venusians). At some point, the Martians, victorious over mere human beings, will have to deal with the greater abilities of the human-created Venusians, or vice versa. Still, even if it's a harsh universe, we come to the same kind of conclusion about the imperative of creative, self-directed evolution. For we had best be prepared to meet it coming from "out there," or else (as Haldane might say) suffer the fate of the first Venusians, or indeed of earthly human life, for our folly.

Flammarion presents what is probably just the prettier side of the same coin. One might have thought that to the scientific mind *any* life beyond Earth would be interesting enough—surely any intelligent life. But the alien forms Flammarion imagines all have some wonderful advantage over mere human beings, so wonderful as to make us look pretty grubby by comparison, "like coarse animals . . . nailed to the soil like any vulgar caterpillar."[147] Should we not then aspire to be more like those superior alien beings? Such evolutionary fitness as human beings might exhibit is relative only to the conditions of our place and time, and perhaps (as Haldane would suggest) we are not suited well enough even for that. We should not expect to persist into the far future, or on worlds beyond our own, without becoming alien to what we are now. But here again the key point is that this change counts as progress. Haldane's Venusians are clearly presented as superior to the humans who created them, and against all squeamishness about crustaceans, Bernal would in effect have us will to become Flammarion's aliens "gifted with multiple sensibilities, luminous at will, incombustible as asbestos, perhaps immortal."[148]

The grand narrative of material progress and self-overcoming has one final twist that again links the materialist-minded Bernal with Fedorov's faith and Flammarion's spiritualism: material progress is itself something to be overcome. Bernal imagines that the ultimate destiny of intelligence may be to resolve itself into light. Whatever that might mean, how different is it from imagining spirits that might be communicated with or resurrected into bodies? Matter can be raised up into life and life raised up into intelligence; why should there not be further extensions of the sequence, however beyond our comprehension they might be? The progress of dehumanization runs from vile bodies to healthy bodies to redesigned bodies to no bodies at all.

THE PARADOX OF PROGRESS

This last transformation into the luminous, if not the numinous, raises in the most acute form a problem has become increasingly obvious as we have proceeded through these lines of thought. In the lead-up to one of his most widely quoted aphorisms, philosopher George Santayana says,

> Progress, far from consisting in change, depends on retentiveness. When change is absolute there remains no being to improve and no direction is set for possible improvement; and when experience is not retained, as among savages, infancy is perpetual. Those who cannot remember the past are condemned to repeat it.[149]

As Santayana suggested, the kind of "absolute" change in the human being imagined by Bernal and Haldane, along with today's transhumanists, really precludes the use of the term "progress." It becomes harder and harder for our authors to imagine what will be retained, hence where change will start from. And if the rate of change is accelerating, that simply means we are headed the more rapidly from one unknown to another, while the recognizable old standards for judging whether the changes are progressive are overthrown along with our humanity.

In today's world, a vision of progress like that laid out by Condorcet remains very much alive. The easing of human life through universal education, reduction in disease, increased sanitation, improved agricultural productivity, and a rising material standard of living is an established fact for much of the world, and the main questions involve how most rapidly and "sustainably" to extend these benefits more widely and how to improve upon what we already have. Likewise, we take increased life expectancy for granted, and worry only about continuing a well-established trend.

Beyond that, even though the thinkers we have examined did not anticipate some of the technological advances that today's transhumanists hang their hopes on—none foresaw the rise of digital computing, for example—some elements of the dehumanization they envisioned are already in place around us. Genetic engineering means that we would not necessarily require generations of careful breeding to create our Venusians. The "conquest of space" is in principle at least an established fact, and if the prospects for space colonies, planetary exploration, and interstellar travel still seem distant, that is less because of what we don't know or can't do than because of how we choose to arrange our funding priorities. Increasingly sophisticated and intimate man-machine interfaces are being developed; we are seeing impressive, if admittedly still early, advances in artificial ears, eyes, and limbs.[150] We may not yet have the organ that Bernal imagined for detecting radio waves, but we do have t-shirts that can display the presence of wi-fi signals.[151]

It is not just what we *do* that links us with the authors we have looked at, but what we *expect*. We don't yet know there is alien life, let alone intelligence, but the idea is widely accepted by scientist and layman alike—if not always for the same reasons. It is likewise a commonplace that we live in a world with an accelerating rate of change.

We might not yet normally place these ideas and achievements within a framework of efforts to overcome the merely human—but they are there to be placed. The eclipse of man is underway. However amazing our present might look from the perspective of a not-so-distant past, there remain those who look down on the human because they can imagine something far better, whether it involves immortality or resurrection of the dead or brains transplanted into machines.[152] Even if, as Bernal warns, we should also be wary of thinking that the future is going to work out just as we envision it today, it would certainly be the height of folly to assume whatever in these visions has not yet happened could never happen. Some (like Malthus) would have said that what we have today is impossible.

The main home for the hopes and fears that define the eclipse of man as we have examined it from the past may be transhumanism, but as we will see in the chapters that follow they are also at work elsewhere—including in the scientific search for extraterrestrial intelligence, to which we turn next. The general public is fascinated by hostile alien invaders. The scientists who look for extraterrestrials are fascinated by contact with advanced, benevolent intelligence. Some transhumanists would be surprised if there are any aliens at all. All these prospects are working out the consequences of ideas about human-alien relations that we have seen in this chapter. The differences among them are not so great as they might first appear.

CHAPTER TWO

Discovering Inhumanity

PROLOGUE: ONLY CONNECT

WHEN SHE accepted a postdoc position to be part of the team decoding the first message ever received from extraterrestrials, Camille never expected that the effort would occupy the better part of her career—would really *be* her career. That was actually the third surprise about the message. The first was that it came in on a tightly focused, extremely powerful beam of modulated UV light, when most of those engaged in the search for extraterrestrial intelligence (SETI) still worked on radio telescopes detecting microwave radiation. At first, the speculation was that whoever was sending the message must be quite technologically advanced to have lasers on a scale that humans were just beginning to think about. But the intensive study of the alien sun that followed showed it to be far more active than Earth's sun; the frequency and intensity of solar storms that Earth astronomers inferred from the data would have made radio communication on their planet so unreliable as to be nearly worthless. Unlike humans, they had probably started and stayed with light as they developed long-distance communication.

The second surprise was that the message was not very user-friendly. It was at least pretty clearly divided into "words," and statistical analyses of their frequency looked a whole lot like what you got from similar analyses of human documents. But beyond that, it was not clear the aliens had considered the audience. Along with everybody else interested in SETI, Camille had given a good deal of thought to how she would design a message that started simple and moved on to more complex topics. At first, the assumption was that the easy stuff had been lost at the undetected beginning of the transmission. After nine months—a dauntingly long letter!—it was clear the message was repeating with no obvious primer at the start. That assumption had to be put aside.

Of course, at the beginning *everybody* was interested. The discovery galvanized and monopolized media attention at least as much as Sputnik and the moon landing. Like most others in her field, deep in her heart Camille had thought that "first contact" would be . . . well, like a revelation from on high. It would change everything. And the fuss at the beginning had done nothing to dissuade her. Interviews, op-eds, news analysis persisted for months. There were two "instant books" on the market within weeks; a couple of the senior people on Camille's team were still living down some of the things they were quoted as saying in those early days. As transmitted, the signal was invisible to the human eye, but it was tuned down, analogized and transformed, mixed and remixed by countless artists in visual and audio forms, bits and pieces of it showing up in popular music and on t-shirts. Camille's prior interest in SETI put her well ahead of the game; there was a huge "catchup" increase in interest in astronomy, optical engineering, linguistics, mathematics, and even astrobiology now that it was a real discipline. A predictable glut of Ph.D.'s in all these areas followed. Even the shifts in government funding for sciences couldn't produce ways to employ them all, although that's what funded Camille's early years. The space program was reinvigorated, two sports teams abandoned Native American names in favor of "ALIENS." Once the message was complete, three "unauthorized" translations were out within months. The only one that didn't make it into print didn't have to, as the author "proved" on his website that the message was none other than the King James Bible.

At the time, Camille had been too busy and too much on the inside to appreciate fully how the message was like a great rock dropped into a small pond. Ripples spread widely, reflected back on each other, interfered and formed a complex pattern. But as time went by and the message remained enigmatic, the disturbance in society at large faded; life returned pretty much to normal. A rump group of enthusiasts stayed focused on translating the message, and some people spent what seemed like all their time trying to show the whole thing was a hoax ("They say the message is transmitted on a light-beam, but YOU CAN'T SEE IT!"), but the vast majority of the world's population went on exactly as before. That humans now knew there was intelligence "out there" became a historical fact among historical facts, part of the background against which the human drama continued to play pretty much as usual. Despite her own dedication to the project, Camille concluded that the discovery really was not, as some had claimed it would be, "the most significant event in the modern history of mankind"[1] or still less "likely . . . the most earthshaking event in human history"[2] or "perhaps *the* greatest discovery in scientific history."[3] It certainly didn't "change everything,"[4] or "cause the most dramatic shift in the status of our human species that has ever occurred in history,"[5] which Camille came to count as her fourth surprise. It was not yet her last.

As the academic work of decoding went on and on and on, various schools of thought formed, competing journals were established on the basis of divergent assumptions, there were conferences you went to and those you didn't. Camille did her best to be a uniter (consistent with keeping favor with her funders) and thank goodness the factions never lost contact with each other entirely, so when her final breakthrough came, nobody ended up a dissident prisoner of his own previous assumptions.

No, the last surprise was not that everybody essentially agreed about what the message said, it was rather the message itself. It proved to be a comprehensive history of the aliens' world (the science parts proved key to the translation, of course). They wanted us to know where they had been, because they were concerned about where they were going. Their admitted flaws and imperfections were becoming increasingly dangerous, they thought, as their powers over their world (they didn't seem to have a concept of "nature") grew. Camille was the one who had first understood how the message began and ended, and it still chilled her when she looked back on it: "Can you help us?"[6]

WE SAW in the previous chapter how human re-engineering is related to ideas about space exploration—and has been at least since Winwood Reade's popular 1872 book—and to speculation about alien life in space. The link continues to be significant today in the writings of prominent transhumanists, but with a new twist. Ray Kurzweil has concluded that "it is likely (although not certain)" that there are no alien civilizations.[7] Nick Bostrom has written that "in the search for extraterrestrial life, no news is good news. It promises a potentially great future for humanity."[8]

Why are transhumanists invested in the idea that we are alone in the universe? Kurzweil and Bostrom each draw their conclusions following out a similar logic. Stars and galaxies are, in comparison with the time it took for life and civilization to evolve on Earth, very old. There would have been plenty of time before we arrived on the scene for an alien civilization to have come into being. So such an alien civilization could be thousands or tens of thousands or hundreds of thousands of years "ahead" of us in terms of its science and technology. Even with the great distances involved, it has been estimated that an alien civilization—one more advanced than ours, but not unimaginably more advanced—could colonize our galaxy in perhaps 60 to 300 million years.[9] From a cosmic perspective, that is a relatively short time. If an alien civilization were to evolve in the way some transhumanists believe we will evolve, achieving great powers to manipulate matter and travel great distances, then surely it would have left its mark on the cosmos.[10] Yet we see no evidence for such a thing. So, as the great physicist Enrico Fermi is said to have asked, where are they?[11] Kurzweil and Bostrom plainly doubt the aliens are there to be found. To understand more fully transhumanist hopes and fears about alien civilizations, it is necessary to take a few steps back, and recall some of the earlier links between aliens and the eclipse of man.

As we saw in Chapter One, human space travel has been proposed as a way to solve the supposed Malthusian consequences of any Condorcet-like vision of material progress. If ever more people are going to be leading longer and wealthier lives, then they will require ever more of the finite resources upon which those lives depend. If we cannot do ever more with ever less, the argument goes, then either human civilization will come crashing down or the resource base will need to be expanded. Space travel, exploration, and settlement, however technically formidable, is conceptually a familiar solution—especially for a civilization, like ours, that was profoundly shaped by its own history of exploration, colonization, and expanding frontiers. So space exploration can seem like a way to protect and extend humanity. However, the genuinely "alien" conditions that prevail in space and on other worlds put a premium on imagining intelligent beings better suited to these environments. Having learned from Darwin that evolutionary diversity is a product of changing environmental circumstances over time, we can readily imagine how evolution might be deliberately helped along to our own advantage. Think of Haldane's humans, bred to select for qualities conducive to survival on Venus, or Bernal's attempt to imagine mechanical beings built for hostile extraterrestrial conditions.

So some human beings *become* aliens to explore and settle new worlds. Bernal acknowledges that this result may be problematic from the point of view of any who choose to remain merely human. Perhaps there is some further reflection of that problem in the complete equanimity with which Haldane's Venusians report the end of humanity on Earth—the humanity that had created them and made Venus habitable for them.

But there is an additional problem. If we can expect to become alien and indifferent to *ourselves*, what if the universe is *not* waiting for us to enliven it? What if Flammarion is correct, and life establishes itself at the slightest opportunity? If there is alien life, then why should there not be alien intelligence? And if there is alien intelligence, why would it not eventually find itself facing the same limits and opportunities that, based on Malthusian assumptions, would drive us into space? If, as Haldane saw, *we* could become alien invaders faced with an imperative to destroy or be destroyed, why shouldn't extraterrestrials behave in exactly the same way?

Of course, based on complete ignorance, we can say anything we wish about alien motives and abilities, a freedom much employed by those who write both fictional and speculative non-fiction works on this topic. So it is not hard to find grounds for happier outcomes, starting from skepticism about the Malthusian dilemma itself. But for our present purposes, the significant point is this: concerns about hostile aliens do not have to arise from commonly identified factors like primitive xenophobia or Cold War paranoia. They do not have to depend on any quirk of human history or psychology. As we have seen, when we imagine dangerous aliens we are imagining beings that are acting no better and no worse than we would act if we fulfill the hopes articulated by today's transhumanists.[12]

Starting from this worldview, then, it is not immediately obvious that contact with aliens would be a good idea. Leave aside all prospects for tragic cultural misunderstandings; on the essential point we may understand each other only too well: they may not come in peace. Unless we are confident we would have the upper hand in the relationship, it might well be thought best to lay low, cosmically speaking.

MAKING FRIENDS WITH ALIENS

After World War II, advances in technology made it possible for the first time to think seriously about what it would mean to communicate across interstellar distances. Thanks to science fiction, the theme of hostile aliens was by then well established in the popular culture. So those who were advocating the scientific search for extraterrestrial intelligence (SETI) had a problem: why would we want to have contact with unfriendly aliens? The first efforts at SETI, led by the American astronomer Frank Drake in 1960, were just a matter of listening with a radio telescope for what was hoped might be the background chatter of alien intelligence—the interstellar equivalent of tuning a radio to eavesdrop on conversations among truckers, or police, or ham-radio users. But those first efforts were rather quickly followed by deliberate attempts to send out contact signals of our own, over and above the radio and television broadcast signals that were already leaking into outer space.

Fortunately for the SETI pioneers, there was a readily available reason *not* to worry about giving ourselves away. By the 1960s, the prospects for other intelligent life in our own solar system were looking bleak, and the distance to the nearest stars provided a comforting buffer. Messages traveling at the speed of light would take more than four years to reach even just the star nearest to our sun. Any back-and-forth communication given this limit would be difficult enough—likely a project of generations, given that our part of the galaxy is not very densely populated by stars in comparison to some other parts. Visits in person, including marauding fleets of star cruisers, seemed, to say the least, highly implausible. So the scenario of aliens exploiting our world for resources to solve their Malthusian problems did not look plausible. We could reach out safely.

There was only one catch. Many of the supporters of SETI believed that any contact we would make would be with aliens more scientifically and technologically advanced than we. There is a simple logic to this familiar belief. Our ability to send and receive signals at interstellar distances is still new and remains quite limited. It would be impossible for us to detect signals from anybody much *less* advanced than we ourselves, since even to detect and distinguish the kinds of signals that we could send out would be difficult or impossible for us. Furthermore, the existence of human intelligence at all on Earth, let alone human intelligence with the technology to begin to contemplate interstellar communication is, cosmically speaking, a very, very recent event indeed. Since on the cosmic clock there has been ample time for life and intelligence to have developed elsewhere long before humanity even began to emerge on the scene, at whatever point we might stand on some general scale of abilities and intelligence, any beings we might contact are likely to be well above it.

But this argument, made by sober Ph.D.'s who did everything they could to distinguish themselves from those who believed in "flying saucers" and (later) alien abductions, in fact raises troubling questions. How confident ought we to be that our understanding of nature and the technologically possible is sufficiently definitive as to preclude practical interstellar travel? Couldn't very advanced aliens know how to do things that would look impossible to us, just as some of the things we can do would seem impossible to primitive men? We are regularly told that the universe is a strange and surprising place—might it be even stranger and more surprising than we now suppose? Even if decades have gone by without aliens appearing on our doorstep, perhaps we should not be confident about what tomorrow holds.

So it proves necessary to attack the very premise of hostile aliens, to turn them from cosmic pirates to cosmic philanthropists. And that effort requires following up on a different aspect of the eclipse of man. For along with this positive view of aliens comes a very dark picture of human beings. As foreshadowed by nearly all of the post-Condorcet thinkers we discussed in Chapter One, the more that alien beings look like our saviors, the less we look worth saving.

HUMAN INSIGNIFICANCE

There are three aspects of the SETI effort to put human beings in their proper—that is, insignificant—place in the cosmic scheme of things. The first is the claim that mankind does not occupy a privileged place in the universe; the second suggests that we should expect aliens to be our moral superiors; the third attempts to show how aliens might help us to overcome our own tendencies toward self-destruction. These three lines of argument combine to suggest that, because anything we can do they can do better, we would be best off following any lead that aliens were willing to give.

The first claim, about the smallness of humanity in the face of the infinite possibilities of cosmic life, is sometimes called the "assumption of mediocrity." Longtime SETI researcher Seth Shostak put it this way: "There's nothing remarkable, nothing the least bit special, about our cosmic situation."[13] Just as we know that there are lots of galaxies like ours and huge numbers of stars like ours, so we have every reason to believe that there are lots of planets like ours. (So far, planet-hunting efforts have revealed lots of planets, but few like ours.) And if there are lots of planets, there are going to be many opportunities for life to arise, as Flammarion argued. (This, too, is a matter about which we have no clear-cut evidence.) And given those huge numbers of stars and planets, then if there is lots of life there will have been ample opportunities for intelligence to arise in at least some cases (or maybe even all). In a universe teeming with life, we just aren't that special.

The assumption of mediocrity is sometimes also called the "Copernican principle," a nod to the "decentering" of humanity that began with the discovery by Nicolaus Copernicus that the Earth is not at the center of the cosmos. Darwin too is sometimes said to have made a contribution in rejecting the necessity for any special creation to explain the existence of humanity. As Shostak tells the story, "These events caused us to lose our central role in the physical and biological realms."[14] If we were to discover evidence of alien intelligence, he believes, it would "surely deliver a roundhouse punch to any remaining hubris, such as the belief that we are intellectually, culturally, or morally superior."[15] *

As yet we have no evidence to suggest how common extraterrestrial life of any kind is, let alone intelligent life. For the time being, then, as Carl Sagan, the great popularizer of astronomy and SETI, acknowledged, "the application of this method [i.e., the assumption of mediocrity] to areas where we have little knowledge is essentially an act of faith."[16] And it could remain a matter of faith indefinitely. Even if we were to imagine some arbitrarily large number of technologically sophisticated alien civilizations—and given the hundred-plus billion stars in our galaxy and the hundred-plus billion galaxies in the universe, there could be a staggeringly high number of intelligent alien civilizations—they might still be spread so thinly that, short of miraculous-seeming technology, contact among them would be unlikely. That would be cosmic quarantine on the grandest scale.

Still, should we ever make contact with aliens, vindication of the assumption of mediocrity will supposedly teach human beings important moral lessons, as Shostak's use of the word "hubris" suggests. For one thing, the knowledge that we are not alone would teach us something very important about our common humanity. "Just learning of the existence of other civilizations in space," Frank Drake wrote, "could catapult nations into a new unity of purpose. Indeed, the search activity itself reminds us that the differences among nations are as nothing compared with the differences among worlds."[17] Sagan apparently shared this belief: "There will be a deparochialization of

---

* It ought to be noted, simply for the sake of intellectual honesty, that the traditional cosmology that put Earth at the center of things did so not out of a hubristic sense that we were the most important beings, but rather because, in the words of Alexandre Koyré, the center represented the bottom of a hierarchy "rising from the dark, heavy and imperfect earth to the higher and higher perfection of the stars and heavenly spheres." By contrast, the idea that we are not unique could arguably represent a significant elevation of human status—for example, it allows Carl Sagan to suggest that we are "star-stuff." (Alexandre Koyré, *From the Closed World to the Infinite Universe* [Baltimore: The John Hopkins Press, 1957], 2.)

the way we view the cosmos and ourselves. There will be a new perspective on the differences we perceive among ourselves once we grasp the enormous differences that will exist between us and beings elsewhere—beings with whom we have nonetheless a serious commonality of intellectual interest."[18] These are very edifying lessons, to be sure, even if one can conceive of a "human family" (an idea whose roots are to be found in the Bible and ancient Greek thought) without SETI. It is quite easy for what is presented as a future prospect to elide into a cosmopolitan norm that the authors almost certainly believe should be guiding our behavior even before any discovery of aliens.

However, encouraging a sense of brotherhood only among human beings does not have to be the end of the story. In exchange for the loss of our cosmic specialness "we will gain something perhaps more desirable," according to physicist Gerald Feinberg and biochemist Robert Shapiro. "Earthlife will be a part of the much more encompassing phenomenon of Universelife. Just as people gradually extend their vision from self-preoccupation as children to identification with a wider community as adults, so in the coming centuries the human species may prefer to be a local representative of Universelife rather than exist almost alone in a vast and sterile void. If so, this shift in attitude may prove to be the most important outcome of a successful search for extraterrestrial life."[19]

Voltaire famously asserted that nothing human was alien to him; Feinberg and Shapiro seem to be suggesting that the yet more mature view would be that nothing alien is alien to us. They hope we may one day be able to aspire to an identity beyond the human in a universe enlivened by many kinds of beings other than ourselves. Just as Flammarion's picture of the wonder and beauty of alien life on alien worlds highlights all that we might find problematic about being human, if we have sufficient faith in the assumption of mediocrity, we could adopt a point of view that weakens our attachment to the merely human, and in this way further the eclipse of man.

HUMAN FOLLY AND ALIEN ASSISTANCE

For many SETI authors, that detachment from the human can hardly come too soon. Along with many others in the second half of the twentieth century, they looked at the world and saw it on the brink of destruction. In SETI circles the discussion has usually been couched in terms of the "L factor," meaning the lifetime of technological civilizations capable of sending and receiving an interstellar signal. The shorter L is, for us or for any aliens, the less likely it is that any two such civilizations will be sufficiently close in time and space to allow communication to take place. Over the years various perils have been used to suggest why we should not be confident that human beings, or at least our present level of technological development, will be around much longer. If there is nothing special about humanity, and if humanity seems bent on self-destruction, then perhaps it is a general rule that technological civilizations are short lived, as the following characteristically apocalyptic quotes from various sources suggest:

> There is a sober possibility that *L* for Earth will be measured in decades. On the other hand, it is possible that international political differences will be permanently settled, and that *L* may be measured in geological time. It is conceivable that on other worlds, the resolution of national conflicts and the establishment of planetary governments are accomplished before weapons of mass destruction become available.[20]

> You might also say that the present very modest trends to reverse the nuclear arms race [in the late 1980s] are an indication that the "L" variable . . . might be larger. On the other hand, the other catastrophes that have come to the fore since then—global warming, ozone depletion, nuclear winter—work in the other direction.[21]

> The mushroom cloud rising above Hiroshima and Nagasaki clearly warned humanity of our potential extinction twelve years before Sputnik soared into space and pointed to our next step in evolution. The planet has managed to avoid nuclear destruction over the past fifty years, but ecological hazards remain a clear threat. Some scientists studying the rapid extinction of terrestrial species today compare it with the time of the great dinosaurs.[22]

> The intelligence and dexterity required to build radios are useful for other purposes that have been the hallmark of our species for much longer than have radios, such as devices for mass killing and means of environmental destruction. We are now so potent at doing both that we are gradually stewing in our civilization's juices. We may not enjoy the luxury of an end by slow stewing. Half a dozen countries have the means to bring us to a quick end, and other countries are eagerly seeking to acquire those means. The wisdom of some past leaders of bomb-possessing nations, or of some present leaders of bomb-seeking nations, does not encourage me to believe that the Earth will have humans and their radios much longer.[23]

> Unfortunately, because the human race is squandering its natural capital at an unsustainable rate, it seems at least as likely that terrestrial collapse rather than extraterrestrial expansion awaits us in the next two centuries.[24]

All these challenges mean that we live at an unusual moment, a time of testing, or a "great filter," as one author has called it.[25] If we pass through successfully, there is hope: we come out on the other side in a brave new world. And that is just where contact with aliens might help us. Any contact we make almost has to be with a longer-L civilization than our own, which in turn means that it is likely to have surmounted just the existential problems that threaten to keep our "L" short. So might they not be able to give us a leg up?

Sagan's writings sometimes show a nuanced understanding of this possibility. He recognizes that there is a problem in expecting to hear from a civilization advanced "vastly" ahead of us; we may no more be able to appreciate their existence than "an ant performing his anty labors by the side of a suburban swimming pool has a profound sense of the presence of a superior technical civilization all around him."[26] (And, we might wonder, would that superior civilization be any more interested in us, or well intentioned toward us, than the suburbanite would be toward the ant?) Hence, most likely our first communication will be with "civilizations only somewhat in our future."[27] Sagan notes that "it has been suggested that the contents of the initial message received will contain instructions for avoiding our own self-destruction, a possibly common fate of societies shortly after they reach the technical phase."[28] Given the likely differences between humanity and the aliens we might meet, he concedes that they might not be able to send us information about "stabilizing societies," but "it is a possibility not worth ignoring."[29]

Why isn't Sagan certain we could be helped? Here another important SETI principle must be mentioned: anti-anthropocentrism. The assumption of mediocrity—the idea that there is nothing special about life or even intelligent life—is not taken to mean that the particular form we take, and the particular way our intelligence builds a human civilization, is not special. Indeed, it would be a great surprise to most SETI scientists if we *were* mediocre in the sense that the alien life we found were somehow to resemble us physically. Notwithstanding popular ideas about "little green men" and other humanoid aliens, most scientist-speculators have long envisioned a wild diversity of alien life.[30] On the other hand, the possibility that they could be, like us, carbon-based life forms, is often taken seriously. We would hardly expect them to communicate with familiar sounds or images, but we do hope they have technologies like ours and share sufficiently common scientific or mathematical knowledge to send an intelligible signal on that basis—although even this possibility has been disputed.[31] In any case, were we to be able to decode a signal, that would at least suggest some commonality of thought processes—but then we might never be sure that we were not missing some (to them) crucial nuance. It might make a difference to the way we understand a message if we knew it was the alien equivalent of a sixth-grade science project, the product of a marketing firm looking for customers, or of a government agency tasked to identify and mitigate potential security threats—none of which possibilities might be immediately obvious to us from the contents of the message itself.

Such potential roadblocks are why Philip Morrison, a physicist who coauthored the first scientific paper explaining how one might do SETI using radio telescopes,[32] had a more sober assessment than Sagan of what might be learned from aliens. Morrison, participating in a U.S.-Soviet conference in 1971, suggested that the impact of an alien message would come less from how it instructs us about questions we are already asking than from the discipline that will arise from the careful deciphering and study of the work of a radically different civilization. He argued—in the process creating some controversy at the conference—that we ought not to expect "some simple telegraphic message like a newspaper."[33] Hence, a signal will have "great impact—but slowly and soberly mediated, transmitted through all those filter devices of scholars who have to interpret and publish a book, and so forth."[34] The "interpretation of the signal will be a social task comparable to that of a very large discipline, or branch of learning."[35] What might we learn from this enterprise? As Morrison put it at a 1972 conference:

> I think the most important thing the message will bring us, if we can finally understand it, will be a description, if one exists at all, of how these beings were able to fashion a world in which they could live, persevere, and maintain something of worth and beauty for a long period of time. Again, we will not be able

to translate it directly and make our institutions like theirs; the circumstance will be too different. But something of it will come through in this way. I think, therefore, that this will be the most important message we could receive. But it will be more of a subtle, long-lasting, complex, debatable effect than a sudden revelation of truth, like letters written in fire in the sky.[36]

IMMORTAL LONGINGS

But among the early SETI advocates Morrison was unusually cautious. There is a more widespread tendency to be optimistic about the ability of aliens to solve at least some of our problems for us. Even Sagan said, "imagine if one day the contents of 100,000 books of a Type II civilization [one that has completely mastered its own solar system] suddenly fluttered through the receivers of our radio telescopes, a kind of Encyclopedia Galactica for children! The rewards of success are inestimable."[37] Drake has been yet more outspoken on this point: "There is probably no quicker route to wisdom than to be the student of more-advanced civilizations."[38] What does he have in mind? "A simple yes or no answer to the question of whether fusion-energy research should be pursued would be worth tens of billions of dollars to the governments of Earth."[39] Or again: "I can only guess what a civilization far more advanced than our own might teach us. . . . But let me share one of my favorite 'what ifs' with you: What if *they* are immortal?"[40]

Drake is suggesting two kinds of gain by contact with intelligent alien life. The first is technical know-how: how to achieve fusion power or immortality. This enthusiastic belief that there might be advanced technological goodies to be got plainly stems from the assumption of mediocrity as applied to technological development. Speaking of aliens as "more advanced" suggests that the timeline along which our knowledge and abilities have developed will be more or less the same elsewhere—that our future will resemble their past. But since our developmental path includes the possibility of being destroyed by our technology, it is not enough to hope that we will get yet more dangerous toys; we will also find the "route to wisdom" that they must have if they were able to surmount the dangers we face. (Of course, this argument overlooks the possibility, highlighted in the prologue to this chapter, that *they* are hoping for assistance in getting beyond difficulties that we cannot even begin to foresee.)

If we look again at the litany of dangers we pose to ourselves, we might conclude that they all point toward some combination of Malthusian scarcity and Darwinian competition. It would only be consistent with the premises of SETI—and not implausible in its own right—to think that these are constraints that any alien life would have to deal with. Here, then, the assumption of mediocrity would properly seem to triumph. But wouldn't anti-anthropocentrism be a safer bet with respect to the *particular* timeline on which such dangers would be faced? When it comes to discoveries and inventions, there are surely some material and intellectual preconditions that suggest the sequence in which things might be discovered or invented; for example, we couldn't have invented rockets without prior experience in combustion. But while such conditions might be necessary, they are not sufficient. It also seems true that there is a high level of contingency with respect to the motives that turn possibilities for discovery and invention into realities.[41] Would we have had atomic weapons in the mid-twentieth century, for example, without the Treaty of Versailles?

Even leaving aside the contingency that seems to characterize many of our scientific and technological achievements, we might notice that our science and technology have developed along the lines that they have because we have the specific human abilities and needs that we do, based, for example, on the shapes of our bodies, the capacities of our senses, the needs that must be satisfied to keep us alive. Add to that the specific raw materials that living on Earth provides to us. All these things shape our particular constellations of abilities, motives, and social structures, which in turn influence scientific and technological developments, and the manner in which they will be useful or dangerous. It is very hard to imagine what it would mean to be made wise by aliens, whose lives, anti-anthropocentrism requires us to believe, may hardly even begin to raise issues that for us require wisdom. To assume that intelligent civilizations all go through similar developmental stages is to adopt an extremely anthropocentric mediocrity, hubristically turning human experience into a cosmic norm.

But Drake himself hints at a kind of solution to this problem when he speaks of becoming immortal. Such a change, if revealed to us, would fundamentally alter what it means to be human. Drake is not being anthropocentric if from the start he is really assuming that "wisdom" means precisely the abandonment of our humanity, and the adoption of the ways of the "advanced" aliens. The point is made with startling directness by physicist-philosopher Paul Davies: "If they practiced anything remotely like a religion, we should surely soon wish to abandon our own and be converted to theirs."[42]

What is perhaps most striking about this assertion is that Davies surely knows that mainstream Christian belief has no problem imagining alien intelligence that would even be subject to salvation.[43] In other words, the reason that we would want to adopt an alien religion is not because all human religions are incapable of dealing with the reality of alien life. Rather, the sole justification would seem to be their presumed civilizational longevity and technological and scientific superiority. In other words, if they are so much more advanced than we are and still have religion, there must be something to it.

The abandonment of central aspects of human civilization is really only to be expected if we are also to abolish hitherto central aspects of the human condition. That may not be such a big deal if we are as spiritually primitive as SETI popularizers seem to imagine, if as Haldane suggested we are an infection hardly worth sterilizing,[44] or if we are, as anthropologist Ashley Montagu once suggested in some remarks about SETI, a cosmic version of "rabies or cancer or cholera."[45] It is true that humans have long abased themselves in the face of Divinity, but Shostak's concern about hubris seems almost quaint in this race to use the authority of science to diminish our sense of what it means to be human.

WHAT'S KEEPING THEM SO LONG?

Let us now return to Fermi's question: where are they? As we have seen, SETI scientists argue that life could have sprung up and developed elsewhere far earlier than it did on Earth, and that other planets might thus have civilizations that were already flourishing and growing in technological capacity over all those eons during which human beings did not even exist. They would have been accumulating knowledge and capacities over what for us is evolutionary time. If they are doing some of the things that the transhumanists predict we will someday be doing—like "reorganizing matter and energy" so as to start turning the cosmos into one big computer—shouldn't we see signs of it underway already?[46] Why haven't we heard from them yet, or detected their work?

Many explanations have been provided to explain the silence of these ancient alien civilizations, but all of them have problems. Perhaps ancient alien civilizations may place young civilizations like our own in a cosmic quarantine for study or other purposes, maybe even leaving behind tools for covert observation.[47] Or perhaps we may just be too primitive for them to bother with.[48] But in a big universe, is there not even one deviant culture that is sufficiently committed and resourceful to want to interact with primitives?[49] Perhaps, instead, at a certain point in their development, technological cultures turn inward, abandoning concern with space.[50] But here again, the suggestion seems to be that there is not even one civilization so taken by spacefaring that it departs from the norm. So perhaps we just cannot hear the aliens because they long ago stopped broadcasting over radio waves—just as we ourselves have in recent decades replaced much of our over-the-air broadcasting with other forms of transmission that do not leak into space.[51] Or perhaps, as some transhumanists expect for our own future, the aliens have given up on physical bodies, "retreating into cyberspace."[52] Yet even then, might they not leave indirect evidence of their existence, such as signs of intensive use of matter and energy to satisfy their need for resources?[53]

Perhaps interstellar travel is far more difficult or costly than we think.[54] But that argument runs up against the suggestion that the development of civilization over evolutionary time scales may produce beings that are simply incomprehensible to us, while we are of as little interest to them as Sagan's ants by the swimming pool. Of course, that in turn means that aliens might have been here all along, and we have been looking for the wrong sorts of evidence. Perhaps instead of searching for big spaceships we should be on the lookout for relatively small probes,[55] or even probes tinier than living cells, or information encoded in the very physics and math of our universe itself, as Sagan imagined in his novel *Contact*.[56] But in a galaxy (or universe?) filled with such life, why would there not be some who would enjoy the challenge of communicating with lower beings just as we imagine how it would be interesting to communicate with animals? There might be, along lines that Sagan has suggested, "a few specialists" among the aliens who study "primitive planetary societies" like our own.[57] They might even leave a message on the Internet.[58] Or perhaps they have been trying to communicate with us all along and we just have not taken the possibility seriously—and we should start rethinking UFOs.

There are two kinds of conclusions that can be drawn from the failure of SETI to find evidence of aliens thus far, although it is very important to be clear from the start that this failure emphatically does not mean there is nothing to be found. However, we can note on the one hand that because we know literally nothing about alien intelligence, the only restriction on our reasoning about it is that our conclusions be logically consistent with our premises—premises that, like the assumption of mediocrity, are themselves as yet more or less without empirical foundations. As a result we have great freedom for imagination, ingenuity, and speculation; the price is that any plausible conclusion can be plausibly countered until there actually is an empirically based science of astrobiology and alien intelligence. On the other hand, it is certainly consistent with a variety of premises to suggest that silence means that advanced alien civilizations who we were counting on to help us may not be as interested in providing that help as we might have hoped. To put it another way, if they want to help, they will do so based on their judgment of the appropriate moment for intervention, which may (given the intelligence gap between us) have little or nothing in common with our hopes.

Frank Drake has come up with a way to accommodate this reluctance and still maintain the possibility of our receiving useful information from alien races. If they achieve something like immortality, he posits, they may become extremely risk-averse. That would explain why they are not interested in exploration, colonization, or contact generally, all of which are risky enterprises. But their "fanatic obsession with safety," Drake theorizes, might lead them to broadcast the secret of immortality.[59] By seducing other species with the very immortality that today's transhumanists so fondly desire, the aliens could minimize any threat from afar.

Does Drake realize how chilling his picture of "the immortals" is? If these beings were immortal, then they could travel to the stars; the times involved would mean nothing to them. But instead of extending and expanding the realm of their experience, their mastery of death leaves them mastered by their fear of death. This is the gift that Drake imagines they would wish to pass on to us. If that counts as "benevolence," it is only by a standard very different from any that poor mortals — who aspire to go *per ardua ad astra* (through advertsity to the stars) — would recognize.

But that is really the biggest question posed by SETI's effort to construct benevolent aliens in the face of the Malthusian-Darwinian argument that they might be our cosmic competition. For is it anything other than gross anthropocentrism to believe that our ideas of benevolence will accord with those of an advanced alien civilization? What makes us so confident that the assumption of mediocrity can be applied to the ethics of alien species? There is an uncertainty at the core of the very notion of alien benevolence. It was explored with brilliance in Arthur C. Clarke's 1953 novel *Childhood's End.* We now turn to that novel to explore how the progress of inhumanity, coming as a gift from aliens, might well be a Trojan horse.

THEY'RE HERE

Childhood's End begins in 1975 near the imagined climax of a space race between the Americans and Soviets; both are out to launch nuclear-powered rockets to the moon. But instead, huge alien ships appear over the major cities of the world. After letting tension build for six days, on the seventh humanity hears from someone calling himself Karellen, Supervisor for Earth. Showing "complete and absolute mastery of human affairs," Karellen announces that all international relations are now under control of his people, who come to be called the Overlords.[60] An unnamed nation tests their control, attacking an Overlord ship with nuclear weapons. Nothing happens; the bomb does not go off and no effort is made to punish the offending nation. "It was a more effective, and more demoralizing, treatment than any punitive expedition could have been."[61]

At the beginning, all contact between the Overlords and earthlings takes place through the secretary-general of the United Nations, Rikki Stormgren, who (not surprisingly) quickly comes to trust Karellen. The Overlords leave many human things alone; they don't care about forms of government or economy, but they are not indifferent to oppression or corruption, even of animals.[62] While it remains okay to kill for food or self-defense, sport killing is out, as the spectators of a bullfight find when they are made to feel the pain of the bull.[63] Or again, the Overlords deprive South Africa of sunlight when the country, which by this point has replaced apartheid with a new policy that oppresses its white minority, is too slow to grant all its citizens equality. In any case, the international system of sovereign nation-states was already dying before the Overlords arrived, and within five years of their arrival a World Federation is in the works.[64]

The Overlords quickly bring unprecedented "security, peace and prosperity" to Earth.[65] Only a few malcontents have any doubts about their essential benevolence or attempt to cling uselessly to the past. But one issue chafes even at those who are happy with the new world: the Overlords will not show themselves to humanity. Karellen promises they will do so in fifty years, after which "we can begin our real work."[66]

Fifty years on, the world has been changed "almost beyond recognition" by the "social engineering" of the Overlords, based on powers of great scope and subtlety.[67] Earth has become a progressive, secular-humanist utopia, truly "One World." Everyone is wealthy, healthy, beautiful, and employed in interesting work that yet allows plenty of leisure time. There is almost no crime. Cities have been rebuilt, and the weather is under control. Continuing education is part of everyone's life. The Overlords have provided aircraft that allow people to globetrot simply to attend a party, and the result is an end of all racial and ethnic divisions. Oral contraception and reliable paternity testing have "swept away the last remnants of Puritan aberration" just as religion itself "vanished like the morning dew" when the Overlords provided access to a time machine that could (selectively) display the past, proving that the world's faiths might have had noble, but certainly not divine, origins.[68] So when the Overlords do reveal themselves, at twice human size and with wings, horns, and barbed tails, humanity quickly gets over the shock that they look like devils.

By the middle of the twenty-first century, the human race achieves "as much happiness as any race can ever know"—a golden age.[69] Only a few ask, "Where do we go from here?"[70] One such is young Jan Rodricks, whose dreams of exploring space are thwarted by the Overlords. They have effectively banned any independent human space travel, and refused to share much of their own obviously greater knowledge and capacities in this area. Through a series of what turn out to be not quite coincidences, he comes to be the only human who knows the location of the Overlord home world. He resolves to visit it. Aware that the Overlords travel at nearly the speed of light, he understands that during a round trip he will subjectively experience as four months long, eighty years will have passed on Earth. He nevertheless stows away on an Overlord starship. As usual, when they learn of Rodricks's disobedience, the Overlords punish no one, but the incident prompts a stern press conference by Karellen announcing formally that the humans will never be allowed access to interstellar space. Humans could never deal with the vastness of the galaxy, and the "forces and powers that lie among the stars—forces beyond anything that you can ever imagine."[71]

Soon enough, however, the human race gets a sense of those forces and powers. All children under ten start to act in strange ways, exhibiting paranormal powers. When this breakthrough occurs, the Overlords must at last announce the truth about their mission. They came not to prevent man from destroying himself with nuclear weapons or other modern technology, but to forestall developing investigations into parapsychology that, had they been followed up, might have loosed a kind of mental cancer into the universe. For humankind turns out to be a race gifted with psychic potentials that even the Overlords do not possess, potentials that make it possible for humans to merge with the Overmind, whose servants the Overlords have been all along. Their job has been to prepare humanity for this merging, to cultivate the proper conditions for it. The children are developing a group mind that will eventually be taken up into the Overmind. All of them are evacuated by the Overlords to Australia, where they become increasingly strange as their minds merge, their psychic powers expand, and they prepare under the tutelage of the Overmind to leave their bodies behind.

In the face of the shocking news that humanity as such has no future, people react variously. Many commit suicide, some live on in decline and desperation. Jan Rodricks returns to Earth from the Overlord home world to find he is literally the last human being left. When the Overlords see that the dangerous moment has arrived for the apotheosis of the children they prepare to depart. Jan offers to stay behind and report to them what happens; as a human he can see things the Overlords are not able to observe. Thus, Jan is witness to the complete destruction of Earth, which ensues as the once-human group mind tests its powers and meets its destiny in the Overmind. Karellen departs the solar system with regret not so much for Earth and humanity as for his own people, who for all their knowledge, brilliance, and power can never achieve this union with their masters, the Overmind.

RETHINKING ALIEN BENEVOLENCE

The issue of the benevolence of alien assistance is a live one throughout *Childhood's End*, far more so than this bare summary might indicate. Certainly the outcome of the story puts this question in high relief. How are we to understand a benevolence that results in the destruction of mankind and the Earth itself? What are we to make of an end so traumatic that (we are told) it echoes *backwards* in time (the Overlords insist that time is more complex than humans understand[72]) such that the physical figure of the Overlords becomes the Devil, the Tempter who just like the Overlords provides all apparently good things as he leads on to destruction?[73] There are times when it is clear that Clarke himself understands some of the problems of Overlord benevolence, and others when he provides hints of deeper issues that he may or may not have been willing to confront himself.[74] The matter of alien benevolence can be discussed in three parts that correspond to the three divisions of the novel: first, the initial subjugation of humanity; second, the golden age; and third, the end of humanity.

From early on there are humans who distrust the motives of the Overlords. That mistrust is given expression by the Freedom League. U.N. Secretary-General Stormgren judges the Freedom League leader, Alexander Wainwright, to be "completely honest, and therefore doubly dangerous."[75] Apparently unknown to Wainwright, his organization contains a secret radical faction. They kidnap Stormgren in order to get his cooperation in finding out more about the Overlords. The Overlords rescue him before he is forced to decide whether to cooperate, but the experience prompts him to start his own clandestine effort to get a look at them. His plan has modest success; Stormgren, a modern Moses, is allowed to catch a glimpse of Karellen's huge backside, presumably tail and all. Understanding now why these devil-like beings would want to conceal their appearance, but confident they will make everything right eventually, he takes his limited knowledge to his grave.

When Wainwright first meets with Stormgren to lodge a formal protest about the plans for the World Federation, he asserts among other things that his group objects to the Overlords because they have deprived humanity of its ability "to control our own lives, under God's guidance."[76] In case we miss the irony here, Stormgren himself points out that many religious leaders support the Overlords' policies, but he concludes privately that "the conflict is a religious one, however it may be disguised."[77] Interestingly, on the same page Clarke has Stormgren note that the "fundamental difference" between him and the Freedom League is that he "had faith in Karellen, and they had not." Clarke seems willing to consider the possibility that the conflict stems from something like religious belief on *both* sides.

Karellen, on the other hand, presents the conflict with the Freedom League more conventionally as religion versus reason and science. People like Wainwright "fear we will overthrow their gods"; science can prove religion wrong, or it can destroy religion by ignoring it.[78] "*All* the world's religions cannot be right, and they know it."[79] Given the way things play out, it looks like Karellen is right on this point; Overlord science does eventually destroy human religion (or nearly so, as we will see) both by active measures (the history viewer) and by the material utopia they provide. And yet Clarke describes Karellen's speech to Stormgren as follows: "His voice was somber now, like a great organ rolling its notes from a high cathedral nave."[80] It would be perfectly consistent with the facts of the story thus far if Clarke were highlighting again that the benevolence of the Overlords is at this stage only a matter of faith. Furthermore, there is the troubling fact that the Overlords are also attempting to conceal the kernel of truth in mankind's *mystical* beliefs, which often find their home in the religions the Overlords have carefully discredited.[81] Given their true purpose in coming to Earth, why should we be confident that the already selective history viewer they loan to mankind is a reliable record of the origins of religion? By standing up for science as he does, Karellen is describing the conflict in a way useful to the deception in which he is engaged.

WHAT DO HUMANS WANT?

The golden age that the Overlords usher in would seem to vindicate their good intentions. The "average man" was grateful to them for being freed from want and war; "mankind had grown to trust them, and to accept without question their superhuman altruism."[82] Few people doubt if it really was altruism. Yet we readers are given ample reason to question the golden age. Recall the carefully formulated way it is described: "mankind had achieved as much happiness as any race can ever know."[83] That does not mean that everyone is happy, that the race as a whole is simply happy, or that happiness is the only or best way to measure the value of life.

For example, there are hints that in various ways this utopia does not quite even live up to its own billing. Religion may have withered, but there is one character who is unwilling to participate in a séance out of what her husband sarcastically labels Talmudic objections.[84] The fact that "Puritan aberrations" have been swept away does not mean that men and women do not still conceal their affairs from each other. The fact that couples marry for defined periods does not eliminate the possibility that a man can become indifferent about his wife, and later regret that indifference.[85] Female beauty is common, but that does not stop jealous impulses when one presumably beautiful woman encounters another yet more beautiful.[86] There are still love affairs, and their failure still creates unhappiness.[87] The security and material well-being provided by the Overlords apparently cannot solve all human problems.

Life under the Overlords is quite clearly not for everybody. We have already seen Jan Rodricks's restless dissatisfaction with the way the Overlords have restricted his life possibilities. More generally, we learn that utopia is dull and that it stifles adventure and creativity in the arts and sciences.[88] (This recalls Bernal's quasi-Freudian views of the pitfalls of a "Melanesian" existence, described in Chapter One.) Inner and outer conflict spur creativity, peaceful satisfaction of desires dulls it. Seeking to counter this problem, some creative types set up a socially engineered island society for artists called "New Athens."[89] But it is hard to believe that Clarke is not gently parodying their efforts. The kind of "conflict" that spurs them includes things like riding bicycles up hills, cooking one's own dinner (in perfectly modern kitchens), or attending endless committee meetings. When an Overlord comes to visit the island, he praises some of its composers for the "great ingenuity" of their work, leaving them "to retire with pleased but vaguely baffled expressions."[90]

If socially engineered creativity is somewhat oxymoronic, that tension at least does not apply to the pacifying social engineering of the Overlords themselves. While they are unabashedly colonial administrators, they highlight two differences between their efforts and analogous human efforts.[91] The Overlord visiting New Athens notes that unlike the British in India, the Overlords have "real motives" and "conscious objectives" for what they are doing.[92] Of course, only after the fact are these motives and objectives made clear to the administered. But humanity could have done nothing to stop them in any case, and that is the second salient difference: the massive disparity in knowledge and power between human and Overlord. When Stormgren accuses Karellen of thinking that might makes right, Karellen does not disagree; he rather points out that with sufficient power of the right kind "might" can be exercised in very subtle, indirect and efficient ways.[93] Might does not have to mean brutality. The Overlords say of their own relationship with the Overmind that "no one of intelligence resents the inevitable"; that is really the bottom-line justification for such benevolence as they have shown by creating a utopia for mankind, which in the end they admit was in part a distraction: "What we did to improve your planet, to raise your standards of living, to bring justice and peace—those things we should have done in any event, once we were forced to intervene in your affairs. But all that vast transformation diverted you from the truth, and therefore helped to serve our purpose."[94]

PERPETUAL PEACE

Once the truth about their mission has come out and the children are about to be evacuated, Karellen announces that "all the hopes and dreams of your race are ended now."[95] He wonders whether it would not be "most merciful, to destroy you—as you yourselves would destroy a mortally wounded pet you loved. But this I cannot do. Your future will be your own to choose in the years that are left to you. It is my hope that humanity will go to its rest in peace, knowing that it has not lived in vain."[96]

Why exactly does the human race have no future? Only young children, under the age of ten, will unite with the Overmind; why can't those who remain continue to reproduce? We are told that many committed suicide in the aftermath of the evacuation of the children; New Athens blows itself up. Others live on with reckless abandon and violence as the global social order quickly deteriorates despite the Overlord power to keep order. But now, their mission nearly complete, the Overlords no longer care. They claim they cannot explain to Jan Rodricks why those who remained did not have more children. He suspects the reasons were "psychological," which is no explanation at all.

In fact, however, their failure to answer this question—are human beings unable or unwilling to procreate?—draws us back starkly to the nature of the Overlords' mission. Their job was to stop humanity from the independent development of paranormal powers and prepare the way for those powers to be used to unite with the Overmind. Independent development of psychic abilities would have made humanity like "a telepathic cancer" in the universe—that is, a threat to the Overmind, even if also to itself.[97] You don't cure a cancer by excising only part of it. A human race that continues to exist *and* knows the truth about its potential would be even more of a competitive danger to the Overmind than a human race that had only halting knowledge of what it could do. And, at any rate, the Overlords seem to have good reason to believe that the children will destroy the world when their rapture comes.

If as we have seen the benevolence of aliens in creating "One World" and a golden age is already problematic, how much more so is it when it eventuates in the end of humanity? Even if, hopeless as it would have been, humanity had attempted to resist either the Overlords or the Overmind, it is difficult to imagine a worse outcome than the one Clarke presents. However, that does not seem to be his view of events. Perhaps he begins from assumptions like Flammarion's; it is in the nature of things that species evolve or face extinction, that worlds are born and die. No matter what else happens, *someday* there will be an end to humanity and the Earth. From this point of view we can understand Clarke's attempt to have us believe that the tragedy of his story is not the fate of the Earth and humanity but the situation of the Overlords, caught in their "evolutionary cul-de-sac," servants of a master they cannot really understand, destined always to be bridesmaids and never brides. The offspring of humanity, on the other hand, are moving on to great and glorious things.

Jan comforts himself by thinking that the Overmind draws "into its being all that the human race had ever achieved."[98] However, it is not even remotely clear that his view is justified. Jan's view implies that the children sequestered in the Australian outback have mental powers of assimilating human history and culture—powers that are never presented explicitly. The Overmind, by taking only children under ten, takes those with the *least* knowledge or experience of what it means to be human and indeed there is some stress put on the fact that the children are becoming something not human at all even before they are assimilated.[99] It is the Overlords who seem intent on preserving remnants of Earth's life and human cultures; they go into a great museum that catalogues the evidence of the worlds they have helped to destroy. Nor would Jan's experience being debriefed by the Overlords on their home world give much comfort about the persistence of human things even if they *did* take older minds. However well educated he is, Jan finds himself unable to answer many of the questions put to him about his own race. These men who have been gifted a golden age are not the deepest reservoirs of human hopes, experiences, dreams, and knowledge.

Karellen's assessment of the situation offers cold comfort: "You have given birth to your successors, and it is your tragedy that you will never understand them. . . . For what you have brought into the world may be utterly alien, it may share none of your desires and hopes, it may look upon your greatest achievements as childish toys—yet it is something wonderful, and you will have created it."[100] But of course humans only "created" what is to come in the bare sense of procreating the necessary conduits for the Overmind.

Perhaps a better argument that the Overlords act benevolently can be found in their justification for barring humanity from the stars. Space is far more immense than human beings imagine. "In challenging it," Karellen says, "you would be like ants attempting to label and classify all the grains of sand in all the deserts of the world. Your race, in its present stage of evolution, cannot face that stupendous challenge. One of my duties has been to protect you from the powers and forces that lie among the stars—forces beyond anything you can ever imagine."[101] Within this universe there is a hierarchy; "as we are above you, so there is something above us, using us for its own purposes. We have never discovered what it is, though we have been its tool for ages and dare not disobey it."[102]

A reminder of the sheer scale of the universe is indeed a useful corrective for human hubris—although surely one could still be sympathetic with any industrious ants who against all odds attempt such a great study of sand. Yet what are we to make of the hierarchy Karellen is suggesting? It sounds almost as if the Biblically inspired vision of God, angels (fallen or otherwise), and man is not so far off the mark after all. Merely from the fact that the Overmind operates through

Overlords, we cannot conclude that it is not omnipotent. And from the fact that it is "conscious of intelligence, everywhere," we might suspect it is omniscient and in some sense omnipresent.[103] Karellen only said that "*all* the world's religions cannot be right"; can we go so far as to say that the Overlords suppressed religions not so much because all of them were false as because some of them pointed to the truth?[104]

ALIEN DO-GOODERS

Like Camille in the prologue to this chapter, most SETI researchers would be surprised if aliens wanted any help from us. But the idea that intelligent alien beings can somehow help us is deeply ingrained in the SETI worldview. The proven existence of aliens alone, it is thought, would have salutary moral effects. Beyond that, there is all the scientific, technological, social, and religious advice that we might get from them. This confidence is based on one aspect of the eclipse of man: its sense of the terrible flaws and insignificance of mankind as it is now. Aliens will have had to find ways around the self-destruction that we fear will be our own fate. They will have found ways to match their great power over nature with great moral responsibility.

Arthur C. Clarke's novel forces us to acknowledge another side of the aspiration to overcome our human flaws. In Haldane's vision of the future, Malthusian and Darwinian forces prompted us to aspire to become the alien invaders, destroyers of worlds. Why wouldn't those same forces be at work everywhere there is life? So between SETI and Haldane we seem to have two very different visions at work. But there is a link between them, which *Childhood's End* suggests can be seen starting from an ambiguity within the meaning of benevolence itself.

First, Clarke knows full well that benevolence has a price even on its own terms; golden ages still have tradeoffs. Access to a galactic version of the Internet may help us solve some of our problems, but it will create new ones. Furthermore, he reminds us that not everything that seems benevolent at first glance really is; the Overlords have their own agenda in creating the golden age—it makes their job of ending humanity easier. This deception raises the question of motive, which will always be a problem with benevolence to the extent that we cannot see into the hearts of others. Is it not troubling that even after the Overlords show themselves to humanity, they are presenting a false front?

But even when motives are pure, benevolence must contend with the fact that people see the world in very different ways. Think about a moral experience not uncommon among us. Should you give the alcoholic panhandler the cash he asks for, or a cup of coffee and a sandwich, or a ride to a shelter, or an appointment at a detox center? Who doubts that one of the latter options is better for the individual than cash? But who would be surprised if the recipient had quite a different view of the subject? So there can be dilemmas of benevolence even when everyone shares a common horizon of human experience and assumptions. Why would the situation be any simpler when we start thinking in terms of all those highly advanced alien races SETI is looking for? As Karellen suggests, Overlord kindness is not really comprehensible to us; it is kindness by a completely alien standard. Clarke allows us to see that even after it has empirical support, there would still be a need for faith in alien benevolence.

The Overlords speculate that they were sent to deal with humanity out of some underlying likenesses between them and us. The hopes of SETI are predicated on the universe being such that we can come to communicate with intelligent beings who are at the same time enough like us to be comprehensible and different enough to be useful. But if, as Karellen suggests, fully in agreement with Carl Sagan, we end up being like ants in comparison to the aliens, then Clarke is correct to be concerned that the universe could contain beings that are very much *not* like us. In Clarke's story the only reason we meet nice guys first is because there is a cosmic hierarchy with a superintending intelligence at work, something remarkably like good old-fashioned providence.

Needless to say, SETI is not built on the fictional metaphysics of *Childhood's End*, but on the accepted metaphysic of modern natural science. Hence SETI rejects the paranormal as a means of communication, just as it rejects flying saucers, just as it would reject Flammarion's spiritualism. Strictly speaking, the universe of SETI is not an orderly, hierarchical cosmos—not a *cosmos* at all in the literal, etymological sense of that term—but a chaos of random interactions of matter and energy without any inherent goal or purpose. Insofar as SETI hopes aliens will lead us to the supposed goods that Clarke's novel portrays, whether world peace and plenty or the overcoming of our humanity, it does so from an entirely different set of premises. So even if we accept that as far as humanity's cosmic fate is concerned, things turn out about as well as one could hope for in *Childhood's End*, we are entitled to wonder if that happy outcome is more likely in Clarke's cosmos or in the materialist universe of modern science, with its Malthusian and Darwinian imperatives.

And yet, perhaps we can begin to reconcile the kinder, gentler aspects of the SETI vision with its tougher underlying assumptions if we note how for Clarke the question of benevolence really becomes a question of power. The Overlords can define benevolence as they wish, and they have the means to get humanity to go along. Ancient alien civilizations would be ancient not because they were good or learned to use their powers wisely, but simply because they learned to use them in the service of their own survival. If might makes right, then all of our dilemmas of alien benevolence disappear; how nice for us if they are so powerful they do not have to be brutal, but if they are brutal (by our primitive standards, of course), then we are left with Karellen's dictum: "No one of intelligence resents the inevitable." This is the functional equivalent of Drake's glib assertion that "There is probably no quicker route to wisdom than to be the student of more-advanced civilizations."

If the law of the jungle is the effectual truth of SETI, then it makes more sense that one supposed moral benefit coming from alien contact, as noted earlier, would be basing our sense of human fraternity on there being a non-human "other" for a radically new Us vs. Them. It also helps explain the transhumanist optimism about the meaning of our failure to find intelligence thus far; in the formula made famous on the TV show *Hill Street Blues,* it means we can "do it to them before they do it to us."

ARE WE OUR OWN OVERLORDS?

Because SETI's contribution to the eclipse of man depends on the contribution of aliens, it may seem that the sort of issues that arise would be unique to that particular way of envisioning the future. But we have already seen more than hints that things are not so simple, especially in Bernal's picture of crustacean-like mechanical monsters with human brains. In the chapters that follow we will see that some of the same issues arise whether the source of our dehumanization is messengers from space or laboratories right here on Earth.

Even before there was any understanding of genetics, people could imagine how careful breeding of human beings could enhance our well-being.[105] When knowledge of genetics started to increase, the eugenics movement was quick to arise, and many thinkers developed even greater confidence that great things could be done to improve humanity. Once genetic engineering comes on the scene, our power to mold ourselves in some new image seems only more likely.

The prospects for and problems with genetic engineering have been widely debated, even though with respect to human beings they are still largely speculative. And yet, there are already those who believe that genetic engineering in and of itself will not be the main route by which to create new and improved human beings. At best, it will be one among the "converging technologies" that together will vastly increase our power over the naturally given, the others being cognitive science, information technology, and nanotechnology. Nanotechnology in particular may open the door to a world where human power quickly reaches the ability to do seemingly everything the laws of nature allow. That is where we turn next.

CHAPTER THREE

Enabling Inhumanity

PROLOGUE: THE VOICE OF EXPERIENCE

Yes, I was what you might call an early adopter when it came to living with nanites, and that's why my memories of the world before and after are still so vivid these many years later—I got prosthetic memory even before you could get the software to edit it. It was like being a kid in a candy shop, and with a big allowance to boot. Not everybody was as willing to experiment as I was, of course, but you more or less found your own level, finding people who were chasing the same kind of dream, at least at the moment you were both chasing it. Not that there weren't already some problems between those of us pushing the envelope and those who thought the world needed more boundaries given all the new possibilities. Old-style democracy was still around at that point, and there was some political pushing and shoving as well as some rioting and sundry nastiness on both sides. But it never worried me much at that stage. I was convinced I knew where things were going; those of us who took the plunge were better off—hell just plain *better*—for having done so, and if it was going to come down to any kind of competition I was confident we would carry the day.

And of course I was right about that. I saw the end coming when the people who wanted to limit or reject the new nano-world started to defend themselves using the same "let us do our own thing" principle that *we* had used to defend transformation! By that time we were perfectly happy to leave them alone as they were pretty much irrelevant anyway. But I admit that I was surprised that that victory was not the end of our problems. Actually, it was more like the beginning. We thought the end of the old society, politics, and economics would free us to be wonderful in whatever ways we wanted. Not exactly. There was a saying in the old world, "when nothing is true, everything is permitted." I don't know about that, but with just a small change it sure works the other way: when everything is permitted, nobody is true. You just could not count on people from one day to the next, or sometimes one moment to the next. There were no serious costs to picking up and going somewhere else, to becoming someone else, even to escaping altogether by getting yourself frozen. Responsibility proved to be pretty rare. It also became perfectly clear that bad guys don't become any less bad when they have more stuff and more power; we tried to deal with that by using "white-hat nanites" to fight "black-hat nanites." That worked okay when the people who invented them stuck to the job, but like I just told you they didn't always. And how could the rest of us blame them? Meanwhile, we also rediscovered that letting everybody do his own thing was not the end of conflict; one person's fondest desire might be exactly what another person most wants to avoid.

So there was even more reason for the like-minded to stick together, cutting themselves off as much as possible from anybody who didn't see it their way. We all told ourselves that we'd still be free to do our own thing because we could join any group whenever we wanted. It wasn't long before there were enough instances of people who joined in bad faith and other kinds of organizational sabotage to bring that phase to an end. By that time the white-hat nanites were being developed by artificial intelligence so we thought we'd be pretty well shielded from outsider lifestyle choices.

Yet once the AI systems understood the logic of what they were being asked to do they realized the benefits of defense in depth and started taking what sure looked like offensive measures—only against the bad guys of course. Or at least the ones who looked bad to us.

So yeah it is a pretty tough world out there just now, but there never was a technology that didn't have some kind of downside, right? Maybe when nano really takes off it'll be different. In the meantime you sprouts may not be quite living the dream but . . . sorry, which siren was that? Another Level Seven incursion? Ok kids, you know where to go. With any luck we can finish this later.

FOR TRANSHUMANISTS, nanotechnology opens all kinds of doors. For starters, it will bring great material and economic benefits. "By making it possible to rearrange atoms effectively," writes the transhumanist philosopher-activist Nick Bostrom, nanotechnology "will enable us to transform coal into diamonds, sand into supercomputers, and to remove pollution from the air and tumors from healthy tissue."[1] Simon Young, another cheerleader for transhumanism, adds that "Eventually, through nanotechnology, vast armies of miniature robot workers manufacturing goods at the molecular level will bring productivity levels through the roof, bringing an end to scarcity and want, poverty, and hunger."[2]

But such changes to the world we inhabit may seem minor when compared to the changes in store for who and what we are. Nanotechnology will be one of the main routes to "the redesign of the human organism," Bostrom writes, suggesting dryly that "once there is both nanotechnology and superintelligence, a very wide range of special applications will follow swiftly." Still, "If we have a choice it seems preferable that superintelligence be developed before advanced nanotechnology, as superintelligence could help reduce the risks of nanotechnology but not vice versa."[3] Meanwhile, the prominent transhumanist Ray Kurzweil points to a specific possibility in the way of building new bodies: smart dust. "In the late twenty-first century, the 'real' world will take on many of the characteristics of the virtual world through the means of nanotechnology swarms," trillions of intelligent networked nanobots that will be able to form into shaped clouds that will simulate anything, including a human body.[4] When we go out, "we will have to select our body, our personality, our environment—so many difficult decisions to make! But don't worry—we'll have intelligent swarms of machines to guide us."[5] (Note how readily Kurzweil slides from using this technology to express our choices to using it to guide our choices.)

The actual definition of the term nanotechnology is somewhat controversial, so it may be safest to begin with the basics. The "nano-" in "nanotechnology" is a prefix used in the metric system, much as a "*milli*meter" is one thousandth of a meter and a "*kilo*meter" is one thousand meters. A *nano*meter is one billionth of a meter, and nanotechnology usually refers to manipulating matter at the scale of around one to a hundred nanometers—that is, at the level of molecules and atoms. It is very hard to imagine things this small. For the sake of comparison, a human hair is about 60,000 to 120,000 nanometers wide; a human red blood cell is about 6,000 to 8,000 nanometers in diameter; a DNA molecule is 2 to 12 nanometers in diameter. If every nanometer in the diameter of a CD or a DVD were expanded to the size of the more familiar millimeter, the disc would be 74 miles across.

For the time being, much mainstream nanotechnology research in universities and corporations is focused on finding uses for "nanoparticles"—very fine particles of some existing material—or experimenting with novel molecular structures. At this very tiny scale, substances can have properties that are not present at larger sizes. Adding nanoparticles to other materials can in turn give those materials new properties. At the moment, applications for nanoparticles are relatively prosaic; one can find them in items like sunscreens, disinfectants, and car parts.[6] These nanoparticles and nanomaterials will surely have many uses in the years ahead, but they are hardly revolutionary. Indeed, there is reason to believe that they have a long history: some materials used in the ancient world, like Damascus steel and some kinds of Roman glass, had special properties that today's scientists attribute to nanoparticles.

But experts are confident that potentially revolutionary applications of nanotechnology are not far off. Mihail Roco, founding director of the U.S. government's National Nanotechnology Initiative, foresees future generations of nanotechnology that go well beyond today's "passive" efforts. Already researchers are developing a "second generation," with active devices such as nanoscale motors that could power nanodevices.[7] According to Roco, third-generation systems of nanodevices could, for example, consist of systems of nanowires in the brain to sense or direct the activity of neurons. Or they could be networked nanoscale robotic devices that assemble themselves into dynamic three-dimensional images for highly realistic telecommunication or virtual realities. A fourth generation of nanotechnology could be, like biological systems, self-assembling, blurring the distinction between living and non-living systems.[8]

By the time we get to the third and fourth generations of nanotechnology, which Roco expects to come into existence over the next two decades, nanotechnology begins to look a great deal like what Nick Bostrom, Simon Young, Ray Kurzweil, and other transhumanists have described. This shared vision of the more advanced possibilities for nanotechnology can be traced back to the work of K. Eric Drexler, whose groundbreaking and influential 1986 book *Engines of Creation* remains one of the most thoughtful treatments of the risks and benefits of such great power.

Drexler, born in Alameda, California in 1955, was as a young man intensely interested in the possibility of establishing colonies in space. He started thinking about nanotechnology in 1976, during his time as an undergraduate at the Massachusetts Institute of Technology, and over the next decade he developed and refined his ideas, presenting them in a scientific journal in 1981[9] and then bringing them before the wider public in *Engines of Creation*. In that book, Drexler imagined the nanotechnology future in terms of "nanomachines" designed to manipulate matter at the molecular scale.[10] These nanomachines, he thought, could be used to manufacture all kinds of goods, including themselves—that is, they would be self-replicating. One nanomachine could first build another, and those two could replicate in turn, until millions or billions were available to assemble molecule by molecule whatever additional product they were designed to produce. As Drexler originally envisioned nanomachines, they could manufacture anything—food, rocket engines, human organs, any consumer product—by building it from the bottom up.[11] * The necessary raw materials could be provided by having other nanomachines sort and disassemble recycled goods, or by using natural resources.[12]

With nanomachines programmed to produce just about anything we could imagine, something like the "replicator" in *Star Trek* becomes possible; Drexler said it "might aptly be called a 'genie machine.'"[13] Nanomachines, he expected, would help in the conquest of space, and natural resources from outer space will provide cheap raw materials and hence "a future of great material abundance."[14] They would circulate in our bloodstreams, looking for signs of disease or decay, forming an artificial immune system.[15] They would repair living cells and even redesign them, for there is plenty of room in a human cell for a nanodevice to set up shop. If atoms were the size of marbles, Drexler writes, a single human cell would be a kilometer across. He

---

* It should be noted that Drexler subsequently abandoned this specific route to nanomanufacturing, has expressed his dislike of the terms "nanobots" and "nanites" that many writers have used to popularize this concept, and has disavowed the notion of "tiny, swarming, intelligent, socializing, conniving things" without making fully clear that this description sounds a good deal like what he once expected, or at least resembles the way he described it. However, the alternative course of technological development that he has laid out in works more recent than *Engines of Creation* does not make a substantial difference in reference to the ultimate promise he sees in nanotechnology. (See Eric Drexler, "Why I Hate Nanobots," March 7, 2009, Metamodern [blog], http://metamodern.com/2009/03/07/i-hate-"nanobots"/.)

estimates that a device useful for doing repairs in the cell would at this scale be about the size of a three-story house, guided by a nanoscale-computer system about the size of a building with a footprint the size of a football field, thirty stories tall.[16] Such repair abilities would open the door to revival after cryonic suspension—the ability to freeze a recently deceased or still-living body to preserve it until technologies are developed to cure or revive the patient. This life-extending technology is the single advance Drexler spends the most time considering in *Engines of Creation*.[17] The central problem of cryonic suspension is that the freezing designed to preserve living or recently deceased tissue actually causes serious damage to cells and tissues. With nanotechnology, people sick or dying or recently dead from illnesses that medicine cannot cure could be frozen, put into what would amount to suspended animation, and thawed out as cures became available. Drexler notes that cryonics need not be limited to the sick: the same freezing and thawing approach could be used by anyone who wants to "time travel" into the future.[18]

Drexler's book sparked the public interest in nanotechnology and inspired many scientists to begin working in the field.[19] But Drexler is a prophet accorded only limited honor in the discipline he did so much to develop and popularize; the U.S. National Nanotechnology Initiative makes no mention of him in its "Nanotechnology Timeline" nor in any of its other official materials.[20] His ideas have been subject to controversy at least since the publication of *Engines of Creation*. The book's strong focus on cryonic suspension and reviving the dead put Drexler in suspect company, and the book's frequent use of the term "imagine" (as in "imagine if . . . ") is also not the sort of thing that many scientists and engineers are likely to be comfortable with, particularly when there is an ongoing and spirited debate about the very possibility of the nanomachines he described.[21] There is also among his critics a sense that his early speculations about the promise, and even more the peril, of nanotechnology had an unhealthy effect on public perceptions, raising hopes and fears in relation to things that might never even come to pass.[22]

Yet his pioneering work remains important whatever the misgivings of his critics. Even if it turns out he did not envision the details of self-replicating nanotechnology correctly, the big-picture issues he raises concerning it remain relevant. For what is he thinking about in *Engines of Creation*? First of all, he is considering the consequences of continued progress in miniaturization in a world where what has already been achieved has had huge impact. Second, he is thinking about the significance of an improved ability to develop useful new materials and customize old materials. Third, he is talking about increasingly fine-grained control over natural processes, the productivity gains that could result from this manipulation, and the results of increasing resource availability. Finally, he is thinking about what all these capacities together might do for our ability to cure or prevent disease—or indeed, to do harm to people—and for our ability to choose freely among a widening set of lifestyle possibilities. In and of themselves, none of these trends he claims to see is controversial. Who does not expect the future to hold more miniaturization, more synthetic materials, more control over nature, more ability to help or harm human health?

So even if it turns out that Drexler was wrong about some specifics of nanotechnology back in 1986, he still identified a *trajectory* of technological development that can be taken seriously. He certainly is not alone in thinking that nanotechnology will change everything. But it could be said that he has thought out its perils as well as its promise more than most. He offers a carefully articulated quasi-philosophy of cosmic history that suggests the *necessity* of developing nanotechnology—a framework that is both useful for transhumanists and contains familiar elements from the eclipse of man. He offers a theory of human personality as well—and it is attractive to transhumanists, too, for it provides a justification for not worrying too much about using nanotechnology to transform our minds and bodies. And he offers a vision of what a world remade by nanotechnology ought to look like. All that makes *Engines of Creation* an unusually comprehensive effort to think about the consequences of a new technology even before it exists. The discussion that follows will not take up the technical challenges to Drexler's vision that have been raised, but will simply accept that something like what he described in *Engines of Creation* could indeed be possible. At the very worst, Drexler's discussion of nanotechnology becomes an imaginative case study for trying to think through the consequences of accelerating technological change, and the discontinuities that it might produce in what it means to be human.

DANGER AND DIVERSITY

Drexler suggests with some justice that his book is not so much advocating nanotechnology as promoting "understanding of nanotechnology and its consequences."[23] For all the benefits that he expects to result from his nanomachine "engines of creation," which he calls replicators, he warns they could also readily be "engines of destruction."[24] "Dangerous replicators could easily be too tough, small, and rapidly spreading to stop," he says. That possibility "has become known as the 'gray goo problem,'" in which uncontrolled replicators literally consume the entire earth in an orgy of exponential growth.[25] (This warning is one of the main reasons Drexler is looked at unfavorably; he is accused of raising implausible frightening scenarios before we know that the nanomachines he describes are even possible.[26]) There are other dangerous possibilities, too. Advanced artificial intelligence (AI), Drexler argues, will facilitate and be facilitated by nanotechnology, and will be able to do in seconds what it would take hundreds of years of human engineering to do. States could achieve "destabilizing breakthroughs" with the ability to mass-produce better versions of existing weapons, or "programmable germs and other nasty novelties."[27] "A bomb can only blast things, but nanomachines and AI systems could be used to infiltrate, seize, change and govern a territory or a world."[28] Under these circumstances, no traditional "balance of power" is enough in international relations; advances could be made in a day that would destabilize any status quo.[29] Domestically, totalitarian states could use nanomachines to infiltrate bodies and minds so as to dominate their populations even more ruthlessly, and to make it yet easier to treat their human subjects as disposable.[30]

For such reasons Drexler acknowledges that "it seems we must guide the technology race or die"; clearly "guide" means direct or channel rather than restrain, since "the force of technological evolution makes a mockery of anti-technology movements: democratic movements for local restraint can only restrain the world's democracies, not the world as a whole."[31] It does not follow from this skepticism about local restraint that Drexler is in favor of world government. It would take a global totalitarian government to stop advanced nanotechnology from being developed; the elimination of liberty would be, he believes, too high a price to pay for safety.

How then are we to maximize the desired outcomes and minimize the dangers? Drexler spends a good deal of time discussing engineering techniques that could be used to reduce risks. Beyond design features, however, Drexler calls for serious efforts at "foresight," which involves asking "three questions. What is *possible*, what is *achievable*, and what is *desirable*?"[32] Laws of nature set limits to the possible, and Drexler makes the bold claim that nanotechnology will allow our power over nature to approach rapidly the limits of the possible.[33] Drexler is aware in principle that as the achievable begins to reach the limits of the possible, foresight will have to focus on decisions about what is desirable. That is clearly a moral or ethical question—how people *should* use this amazing expansion of human power.

We are left with a dilemma: "Our differing dreams spur a quest for a future with room for diversity, while our shared fears spur a quest for a future of safety."[34] People's desires for our world and themselves diverge wildly, and so, as Drexler develops the argument, it becomes clear that he believes liberty and diversity go hand in hand. But liberty and safety do not necessarily go together. Leave people free to do as they wish and they may end up hurting each other; the more you want to keep them safe from each other the more you may need to restrict their liberty. Political philosophy has long had to deal with this dilemma. But in Drexler's case, this already serious problem is made yet more difficult by the underlying logic that he thinks will drive nanotechnology development, which strongly constrains the foresight question of what is desirable, even as it extends the concept of diversity well beyond anything that we are familiar with today. First we will examine that logic, and see its ultimate consequences for what diversity means to Drexler, and then turn to how he believes that radical new diversity can be dealt with.

COMPETITIVE PRESSURE

To understand the logic behind the development of nanotechnology, it is important to see that Drexler is not content to argue for nanotechnology on the grounds of its possibility and desirability alone. He also wants to suggest that because of competitive pressure we—understood in some sense as a national "we" and in some sense as a human "we"—have no choice *but* to develop it. The most immediate reason is international competition; whichever country is the "leading force" in nanotechnology will gain a huge economic and military advantage over everyone else.[35] That means there is every incentive for *somebody* to develop it, and to his credit, Drexler, writing toward the close of the Cold War, clearly prefers that it be the United States.

Competitive pressure allows Drexler to place his nanomachines within a broad sweep of evolutionary development. The philosophy of cosmic history that he offers is reminiscent of the way Winwood Reade expanded on and adapted Darwin's ideas in *The Martyrdom of Man* (discussed in Chapter One). Drexler follows Richard Dawkins, the neo-Darwinian evolutionary biologist, in arguing that evolution is all about the "variation and selection of replicators."[36] The earliest simple replicators based on RNA and DNA evolved into a bewildering variety of forms—giving us life as we observe it. Eventually, the story goes, a kind of replicator we call human beings came along, capable of developing technology. This technology in turn undergoes the same process of evolutionary selection, as do the ideas ("memes") that are part and parcel of its development. Nanomachines are just the latest form of replicator, brought about by human replicators.

We must expect, then, that "deep-rooted principles of evolutionary change will shape the development of nanotechnology, even as the distinction between hardware and life begins to blur."[37] Competitive pressure will drive the evolution of these new replicators as it drives all others; "the global technology race has been accelerating for billions of years. The earthworm's blindness could not block the development of sharp-eyed birds."[38] Unlike natural selection, which depends on randomness, the variation and selection of nanomachines can be directed by the proper application of foresight to serve the ends of intelligence. Yet because competition makes the development of nanotechnology necessary, the question of desirability cannot really be a question of "nanotechnology or not" but must rather be a question of what kind and how. Indeed, Drexler argues that foresight must help us select *against* ideas and ways of thinking that stand in the way of accepting this new form of replicator.

By lumping together many different biological, physical, and mental phenomena under the broad category of "replicators," Drexler assimilates the whole history of technology into the history of life, which in turn makes two kinds of evolution possible: one based on chance and the other guided by intelligence. (This distinction is an important point of contact between Drexler's argument and the arguments of the transhumanists, as we will see in the next chapter.) So when we consider the future of nanotechnology, we are seeing a new stage in an old story. Drexler's proof of concept for self-replicating nanomachines is the existence of the self-replicating organic "protein machines" that are so well established in the biological realm already—that is, all the parts that make up cells and organisms.[39] With nanotechnology, we are just developing a new form of a common phenomenon.

MIND IN MACHINES

Yet there will still be significant discontinuity in the shape of life to come. Nanomachines may lead to the eclipse of man, and into realms of achievement we cannot now comprehend. Drexler envisions "revolutions": remaking our bodies, melding mind and machine, spreading life into space.[40] The range of these possibilities suggests that the "diversity" Drexler discusses will have a very wide scope. Still, he is plainly aware that some people will be troubled by the convergence of minds and machines, with all that it implies about what we are and what we may become: "some feel uncomfortable with the idea that machines underlie our own thinking."[41] To make such people less uncomfortable, Drexler attempts to shift their understanding of "machine" away from just the sort of "picture of gross, clanking metal" that Bernal presented in his crustacean-like brain housing. Drexler would rather we imagine "signals flickering through a shifting weave of neural fibers.... The brain's really machinelike machines are of molecular size."[42] But the further implications of this view of what a machine can look like raises its own issues. Signals passing through fibers could describe either human intelligence or artificial intelligence—another idea that Drexler understands people resist. Yet if the brain is *already* understood as a machine, opposition to artificial intelligence becomes merely "biochauvinist prejudice."[43] Drexler is drawing out the consequences of scientific materialism's view of human beings as sophisticated machines. From this point of view, a new transhuman or posthuman model of humanity is no big deal.

The supposed need to overcome biochauvinist prejudice tells us how Drexler's idea of diversity includes dehumanization as a core element of the nanotechnology future. We see another in his discussion of one of the great payoffs he expects from nanomachines: indefinitely extended life. (Like Condorcet, Drexler is careful not to say immortality.) The argument begins with relatively conventional goals for nanomachines: we will use them to cure disease, detect and repair cellular damage, replace old parts with new. However, should there be anything wrong with the body that nanomachines can't fix, they can still make possible revival after cryonic suspension. Echoing Bernal, the only human part that Drexler really believes needs to be frozen is the brain—since the patterns of the information stored in the molecular machines that are our brains define the meaning of the "I" in the phrase "I will live far longer." To make this case, Drexler takes advantage of the real uncertainty we have about the basis for our personal identity and selfhood. Is it the body? Yet the body changes over time, and the very stuff of the body is in constant flux and not at all what it was not that long ago. So where does it reside? As a materialist, Drexler does not want to believe in a soul—but he comes close. The "I" is a pattern of information, a pattern residing in the brain.

"Nature draws no line between living and nonliving," Drexler writes.[44] Matter is matter, but it can be patterned in different ways, and these patterns make a huge difference; "one simple sum of our parts would resemble hamburger, lacking both mind and life."[45] With regard to that brain, however, Drexler turns out to be something of a dualist:

> A mind and the tissue of its brain are like a novel and the paper of its book. Spilled ink or flood damage may harm the book, making the novel difficult to read. Book repair machines could nonetheless restore physical 'health' by removing the foreign ink or drying and repairing the damaged paper fibers. Such treatments would do nothing for the book's content, however, which in a real sense is nonphysical. If the book were a cheap romance with a moldy plot and empty characters, repairs are needed not on the ink and paper, but on the novel.[46]

We know that the nonphysical information of the book, the novel, can have any variety of physical embodiments and remain the same with respect to the information it contains, whether the book be hardcover, paperback, e-book, or audiobook. Of course, the resurrected patterns of information that reside in the brain could be housed in a familiar body—perhaps even a body that resembles the brain's original body. But why stop there? Why not an unfamiliar body, or any body at all? Indeed, once the *pattern* in the brain becomes the key thing, even the brain itself becomes disposable. In a sense, Drexler allows the argument to advance a step beyond Bernal, for whom already the body was a mere tool of the brain. Why should the pattern of information that truly is each of us reside in a biological brain at all? And why must that pattern reside in only one location at a time?

Drexler's presentation of individual identity as a pattern of information is a symptom of the relatively precarious way that consciousness and our sense of self are situated in scientific materialism.[47] In *Engines of Creation* Drexler does not explore in depth all the implications of the ability to "remake our bodies."[48] While he mentions in passing the possibility of "bizarre" modifications of the body, for the most part he is content with letting his readers assume that a resurrected brain will be placed in something like a familiar body, and that one will lead a familiar life with resurrected friends and family—a scene that provided some inspiration for the prologue to this chapter.[49] The implications are nevertheless clear enough, and suggest another respect in which nano-based diversity will produce inhumanity. Information can be stored, copied, manipulated, and transmitted in all kinds of ways. The idea that "I" am just a pattern abstracts not just from bodily particulars, but from how bodies are embedded in the larger world; or, perhaps more precisely, it assumes that we live in our own heads, and we don't even need to stay there. It wipes away much of how we experience and understand ourselves and our world. (In the next chapter we will see how transhumanists take advantage of the potentially radical consequences that pattern identity has for detaching people from "biochauvinist prejudice.")

NANOPOLITICS AND HUMAN IMPERFECTION

Drexler does his best to argue that the kind of diversity he has in mind—where people can choose to live on starships, to colonize alien worlds, to communicate telepathically, to redesign their bodies, to "time-travel" with cryonics, or even to live the way we do today—is in keeping with the kind of wide, free choice that is considered desirable in liberal democracies and capitalist economies.[50] As remarkable as those examples may appear, Drexler seems to suggest that they fall under the rubric of the "pursuit of Happiness" mentioned in the Declaration of Independence. But will people with such diverse desires be able to live together? Drexler is smart enough to understand that we cannot simply assume that giving people more power to choose individually or collectively how to live will make them more tolerant of those who choose a different way of life; it might make them more insular and less tolerant. So it is at least not self-evidently a good idea to seek a future where more people will have larger differences and more power to fight about them, a world in which people believe themselves even more entitled to do as they please than in many parts of the world today. We might hope that such power would be a recipe for leaving each other alone, for doing as we would be done by, but that is not the lesson of experience.

So what is to be done to make sure that nanotechnology is used responsibly—to govern its use in such a way that the gray goo scenario, for example, never comes to pass? Political, legal, or moral restraints that will help us to make the right decisions about the development of nanotechnology do not have to "start from scratch," Drexler writes.[51] There are many elements of liberal democratic society and politics that can provide the necessary foundations. "The principles of representative government, free speech, due process, the rule of law, and protection of human rights will remain crucial," as will "such diffuse and lively institutions as the free press, the research community, and activist networks."[52] Further, such things as "personal restraint, local action, selective delay, international agreement, unilateral strength, and international cooperation" will all be useful for avoiding fearsome scenarios.[53]

However, such institutions and modes of restraint are only useful in a specific context: when they go along with what Drexler calls "active shields." Active shields are defensive, automated nanotechnologies that act as an immune system, seeking out and destroying hostile nanomachines.[54] In other words, the best defense against bad nanomachines in the wrong hands is for nanotechnology in the right hands always to be one step ahead. If we look more closely at the kinds of measures Drexler advocates for guiding the development of nanotechnology, we can understand why in the end Drexler must place a huge amount of weight on active shields.

Political solutions are at least as imperfect with respect to nanotechnology as they are generally; in particular they are not complete answers to the existence of "power, evil, incompetence, and sloth."[55] Drexler acknowledges that democracies can commit "atrocities" and that they contain evil people.[56] The potential for evil is balanced by the fact that democratic "leaders gain power largely by appearing to uphold conventional ideas of good."[57] He never makes clear just what constitutes "good" or "evil," a fact that suggests that he does not have a lot of hope that some "ethics of nanotechnology" will be very useful as a restraint. That power and evil are relatively intractable problems already suggests why active shields will be necessary.

Sloth and incompetence seem to be more tractable problems, although still not easy to eliminate because "we human beings are by nature stupid and ignorant."[58] But in this area as in so many others, Drexler does not think we have to be content with nature. Indeed, we already know how to cooperate on technical matters to "gain reliability through redundancy."[59] Incompetence can thus be weeded out. He also provides suggestions to "improve our institutions for judging important technical facts."[60] One of these ideas is to create "fact forums" (sometimes also called "science courts"), which would put technical disputes within a quasi-judicial framework of due process, so that clear statements of what is agreed upon, and the parameters of disagreements, can emerge. Drexler does not envision them as policymaking bodies; they seem designed largely to lay out the merits of technical disputes for the public and decision-makers.[61]

Even if foresight can be improved by improved competence, sloth is another matter. It may well be that not everyone will be willing to "meet great challenges with great effort."[62] But it is not necessary that everyone be on board from the start. "It will require only that a growing community of people strive to develop, publicize, and implement workable solutions—and that they have a good and growing measure of success," Drexler writes. "Sloth will not snare everyone's effort. Deadly pseudo-solutions (such as blocking research) will lose the battle of ideas if enough people debunk them. And though we face a great challenge, success will make possible the fulfillment of great dreams."[63]

SHIELDS AND LIMITS

While Drexler suggests that incompetence can be minimized and sloth may be made irrelevant under the right conditions, he wisely does not claim to have a complete solution to the two other problems he names: evil and the abuse of or hunger for power. These are facts of life, and no set of beliefs or framework of laws and institutions can restrain them perfectly. But the great power conferred by nanotechnology, power that Drexler himself suggests could change everything in a day, makes evil and the lust for power particularly dangerous. So moral, legal, institutional, and political restraints are likely to mean little unless they go along with active shields. Moreover, moral, legal, institutional, and political restraints are likely to be in tension with the goals of liberty and diversity in a way that active shields are not. The arguments that Drexler uses against efforts to prohibit nanotechnology generally—local restraint alone is ineffective, global restraint totalitarian—would apply just as much to any effort to restrain some particular form or use of nanotechnology. So moral, political, and legal restraints are either ineffective when non-uniform or dangerous to liberty and diversity when uniform. This dilemma may help explain why Drexler is not very interested in exploring terms like "evil" or conventional understandings of what is good. As different ways of life develop around different uses of nanomachines, the meaning of these terms will be contested. If active shields work as Drexler intends, he can think of them as guarantors of libertarian cultural relativism; it will not matter to one protected enclave what is going on in another.

Now, any society, even one that values liberty, needs some kinds of "active shields" like police, private security, or military forces because moral, political, and legal restraints cannot enforce themselves and not everyone will be equally restrained by them. In a society that loves liberty, these human active shields are there to make sure that (to use the familiar phrase) the liberty of your fist ends just prior to my nose. In a civilized society these old-fashioned active shields are the *ultimate* line of defense; they come into play when all other restraints on behavior fail. These systems are imperfect, so crime and conflict are not always prevented. But the intensity of crime and conflict is limited not just by our active shields, but by some degree of unity, even if limited temporally or geographically, on normative beliefs that restrain conflict, and on the relative difficulty of acquiring tools that would be capable of larger- rather than smaller-scale destruction. So although it is relatively easy to get guns in the United States, for example, most people are not going to use them to commit crimes, and the few that have criminal inclinations will find it progressively harder to get progressively more dangerous weapons.

The future world that Drexler invites us to imagine is one with greater diversity than the one we live in, less normative consensus, and easier access to more dangerous tools. Thus, his active shields are not the last line of defense, they are the *precondition* for creating a "stable, durable peace" while maintaining diversity and liberty in a world where human power increases but human goodness may not.[64] Indeed, to the extent that the active shields work, it might appear that the question of goodness has reduced significance. Ensconced behind an active shield, we can safely follow our own vision of the good without having to worry about anyone else's.

Drexler's willingness in some fashion to confront the problem of evil justifies his claim that he rejects standard utopian fare, which "all too often [has] been impossible and the attempt to achieve it has been disastrous."[65] He wants to present us with "useful dreams"; as we will see shortly, neither *we're running out of resources tomorrow* nor *we can do anything we want forever* is a useful dream.[66] Likewise, it is not useful to think that we need to plan today for everything the future might hold; the "great task of our time" is not so much to build this world of diverse dreams as to guide "life and civilization through this transition" to it.[67]

UTOPIAN ANTI-UTOPIANISM

We have already seen some of Drexler's recommendations for making the nanotech future safer despite the presence of power and evil, and for reducing, or at least minimizing the effects of, incompetence and sloth. While he claims these suggestions can build on existing institutions and ideas, will our system of politics survive in a recognizable way? The future Drexler describes seems likely to fundamentally challenge liberal democracy, which is predicated on checks and balances that allow the interests and ambitions of some, running along their well-established lines, to counter the interests and ambitions of others. What happens when those lines can be extended to the limits of the possible? Certainly the liberal democratic "pursuit of Happiness" is *not* predicated on the degree of human malleability and the consequent radical range of choice that Drexler presents, and the same would be true of rights generally and indeed of any of the hitherto existing "conventional ideas of good."[68]

This discontinuity should be understood carefully. There is ample historical precedent for expanding the sphere of moral concern and political protection to formerly excluded classes of human beings and considering it progress. The future could simply hold more of the more-or-less same. At the very least, our conception of the human (as in human rights) would have to expand, or our understanding of rights-bearers would have to shift from human beings to some other category, such as sentient beings. (The already out-of-favor yet foundational idea of "natural rights" will have even less approval in a world bent on reconstructing nature itself.) But even under these "optimistic" circumstances, Drexler's "useful dreams" are of a world that remains extremely dangerous precisely because of his hope for maximizing choice and minimizing restraint. The libertarian world he looks forward to seems unlikely to make people less assertive about their desires and perceived interests given the increasingly unimaginable benefits of nanotechnology, particularly when, for the "I" that is coming to understand itself as a "pattern" capable of diverse embodiments, the act of choosing becomes the locus of identity. The sphere of moral concern would have to be expanded to encompass all possible definitions of its appropriate scope.

In light of the diversity that Drexler envisions, one might ask, "If people can choose to do as they please, why should they concern themselves if others are choosing differently?" One could think even more creatively, and imagine that out of the settlement of alien worlds will spring a plethora of new nanotechnology-enabled species, alien beings who will have little interest in each other. But Drexler has already admitted that there are two flies in this ointment. The first is the existence of ongoing scarcity. This is a subject on which his position is nuanced, if not somewhat obscure. Drexler's original interest in space and then in nanotechnology grew out of a reaction to the doomsaying projections made famous in the Club of Rome's 1972 report *The Limits to Growth*,[69] which he first read as a young man.[70] In *Engines of Creation*, he utterly rejects the report's projections that we are quickly running out of resources; nanotechnology and human expansion into space will put greater resources under human control.[71] But fundamental laws of nature still create limits under which people will have to live. For example, nature dictates that the expansion of humanity into the galaxy will be limited by the speed of light (at best). That natural limit means that if our population grows exponentially, we will not be able to spread rapidly enough to obtain the resources we need. So while "the spread of life and civilization faces no fixed bound," Drexler writes, "unchecked population growth, with or without long life, would overrun available resources in one or two thousand years at most."[72] Drexler, in other words, believes that Malthus was essentially correct. So scarcity will remain an issue, even if only a cosmic-scale matter of living space, and competition among worlds will not disappear.[73]

The second reason that radical diversity will challenge peaceful coexistence and live-and-let-live values becomes clear in an admission that Drexler makes: "Unless your dreams demand that you dominate everyone else. . . . "[74] Such dreams are hardly unknown, but it is not clear Drexler has fully confronted their significance. As Aristotle knew, people do not turn to crime simply "through being cold or hungry" or become tyrants solely "to get in out of the cold."[75] However much we may prefer a world where there is plenty of heat to one where there is not, from the point of view of Aristotle's realism, the abundance that Drexler imagines may encourage desires "beyond the necessary things" or for "enjoyment that comes with pleasures unaccompanied by pains. . . . The greatest injustices are committed out of excess."[76] Excess almost seems to be the point of the nanotechnology that Drexler describes in *Engines of Creation*.

Because some people have dreams of domination, others will need protection from them. Even if greater choice and less scarcity reduces conflict, there will still be reason for conflict so long as there is ordinary crime and so long as diversity includes those (call them evil if you wish) who find reason to dissent from the orthodoxy of diversity itself. If even only a few want more than they can get, or even a few concern themselves with martial glory or the love of honor, or even a few exhibit pathological evil, the diffident and tolerant many will need to be able to protect themselves, and their best defense may be a good offense.

Since Drexler gives us reason to think that active and competent evil will continue to exist in the future, then while he would like to think that active shields are purely defensive and hence non-threatening, they *are* a threat to anyone who has reason to worry about what is going on behind them. And such concerns arise not only so long as the possibility of evil exists. They will arise also out of the very fact that diverse visions of the good will often lead to divergent understandings of what kinds of behaviors are "safe" or "risky" in the first place.[77] Drexler ignores the complication that in his world of choice among such visions, one group's benefit may be another's risk. Unless "worlds" can be completely isolated one from another, the competitive pressures Drexler counts on to produce progress in developing nanotechnology will not disappear even as the power to impose on others will increase, even as wild diversity decreases the sense of solidarity. Even those who might choose to "opt out" of the lifestyles nanotechnology makes possible will still be dependent on it: for example, if *almost* everybody is willing to let me raise my child without nanotechnology, I will still need an active shield against unrestrained deviants who find my behavior to be child abuse. In general, then, it appears that active shields will have to be very active indeed.

Which raises a vital question. Who will develop and maintain the active shields? Not all of the lifestyle choices Drexler speculates about, or that we might imagine, would in and of themselves point to interest in or competence at this important task, particularly to the extent that they share the desire for ease and comfort that Bernal spoke of as "Melanesian" aspirations. But somebody is going to have to mind the store. One wonders whether Bernal did not see more clearly than Drexler the likely outcome of this dilemma: utter dependence of the many on those relative few (perhaps the committed and non-slothful?) who develop and manage the technology the many depend upon. Diversity must be built on uniformity of technological capacity, freedom on dependence on those guardians who develop and maintain active shields. The shield guardians will not always share the values of those they protect because their focus will have to be on the non-diverse technical demands of their job.

Or perhaps—as indeed seems likely—these shield guardians will not be human at all but precisely those advanced artificial intelligences that Drexler counts on to push so rapidly the achievable to the limits of the possible. Indeed, it is hard to imagine how such an artificial intelligence could even begin to be under mere human control, given that its main strategic and tactical advantage over human development, design, and deployment would be the speed at which it would operate. Either human beings would have to be enhanced to catch up to it, or it would have to have been securely programmed (or otherwise convinced) to be well inclined to those it nominally served. The old question of who will guard the guardians was hard enough to answer in light of the old worldly wisdom that man is a wolf to man. It is far from clear that handing over power to artificial intelligences will represent a solution to this problem.

Drexler is aware that diversity and conflicting values go together. He knows also that there will be "genuine opposition to an open future, based on differing (and often unstated) values and goals."[78] He expects it will be opposed by "the power-hungry, the intolerant idealists, and a handful of sheer people-haters."[79] But despite the conflicts that Malthusian scarcity and Darwinian competition would suggest follow from such diversity, he does what he can to avoid dealing fully with the implications of divergent visions of the good. He holds that such disagreements will be far less important than those related to "differing beliefs regarding matters of fact."[80] Just as successful nanomachines reduce the problem of sloth, so too in the form of active shields are they the key to getting around value disputes.

Unless Drexler believes that there is an objective basis upon which to distinguish among visions of the good, then clarity about the facts will never be enough to settle such disputes. Rather, it seems that Drexler believes that in the present, the force of necessity allows foresight to slight the question of what is desirable, as if the might of technological possibility makes right. In the future, he hopes that active shields will make it unnecessary for there to be any common answers to that question. Yet the significance he gives to competition and scarcity make it implausible that what happens within active shields will not have consequences for relations among the shielded groups, and therefore on the way they are organized internally to meet this challenge. It is all very well to let a thousand flowers bloom, until you discover that some of them are invasive weeds. If it is impossible to isolate the worlds created by nanotechnological diversity from one another, we cannot overlook the significance of conflicting visions of the good among them.

WHAT SHOULD BE DONE?

This matter that Drexler seems not to address fully—the issue of conflict based on divergent ideas of the good—is at the core of Neal Stephenson's portrayal of a nanotechnology future in his dazzling novel *The Diamond Age* (1995). Stephenson builds his story on broadly Drexlerian assumptions about the future of technology, and indeed, Drexler himself has endorsed the novel.[81] But Stephenson is far more interested than Drexler in the question of how diverse visions of a good life might be organized internally in a nanotech future, and how such organized groups will be influenced by their competitive, external relations with those groups that have differing values. This question of relations between inner and outer arises throughout *The Diamond Age*, relevant not just to the way nanotech-based cultures interact, but also to how people interact as individuals.

The world of *The Diamond Age* is, by our present-day standards, pretty magical. A bird's-eye view would note the following: a worldwide computer network (called "the Net," recognizable as roughly equivalent to our Internet) was responsible for the downfall of the nation-state; once the Net made possible secure financial transactions that could not be traced by governments, people stopped paying their taxes.[82] Nanotechnology produces goods plentifully; nanomachines disassemble matter at the molecular level and, normally for a price, programmable "matter compilers" (Drexler's genie machines) can reassemble it into anything that is wanted.[83] This world, Stephenson says, is one in which "nearly anything" has become possible.[84] But the observation that makes Stephenson's book so insightful is stated early on; for as nearly anything has become possible, the "cultural role in deciding what *should* be done with it had become far more important than imagining what *could* be done with it."[85]

In the absence of the nation-state, answers to the question "What should be done?" are provided by a host of culturally distinct tribes, or "phyles." But all phyles are not equally successful, and those inequalities matter a great deal. The largest and perhaps most powerful are New Atlantis, Nippon, Han, and Hindustan. The major phyles have obvious geographic antecedents and still control territory, including islands they fabricated using nanoconstruction off the coast of China and elsewhere. The story of *The Diamond Age* is largely centered on one such island, called New Chusan. It serves as real estate occupied by members of the New Atlantis phyle, which has self-consciously replicated Victorian norms as the best way to deal with the promises and perils of this new world. New Chusan is also the intake processor for the molecular stocks—the raw materials sorted largely out of the atmosphere and seawater—that are passed to the matter compilers through a network of feeder lines (the Feed).[86] Territory on the island's lowland periphery is leased to other less powerful phyles and to "thetes," who belong to no phyle or to small "synthetic" phyles.[87] The thetes constitute a relatively lawless underclass, living for the most part off the limited selection of free goods provided by the matter compilers.

The Diamond Age is by no stretch of the imagination a utopia.[88] Nanotechnology makes it a frighteningly dangerous place, where thugs install "skull guns" in their foreheads that shoot nanoprojectiles capable of turning a body to mush from the inside out. Successful phyles have extensive security measures to protect them from hostile nanotechnology; these Drexlerian active shields consist of things like tiny hunter-killer airborne nanomachines, whose immune-system-like battles with intruding nanotechnology can turn the air gray with dead "mites."[89] New Atlantans also revive Victorian customs useful for self-defense: reception parlors in their homes serve to scan all visitors for hostile nanomachines, and women wear nanotech veils to fight off the same.[90] The whole New Atlantis enclave is surrounded by a grid of floating security pods.[91]

Yet despite all the advances in nanotechnology, the action of the story develops from the fact that in this new world, people (mostly) remain people. Parents want to do well by their children, who remain difficult to raise.[92] Businesses want to make a profit. Society remains "an elongated state of low-intensity warfare," and high-intensity warfare can break out among competing phyles.[93] *The Diamond Age* pictures a world of prejudice, inequality, exploitation, competition, and crime—but also a world of nobility, self-sacrifice, self-discipline, and virtue. In other words, in its moral fundamentals it is a world very much like ours. Although Stephenson is not blind to the potential for the progress of dehumanization inherent in Drexler's promises, his achievement is to think through the impact of nanotechnology and other new technologies in relation to perennial *human* possibilities.

Stephenson observes how nanomachines, by opening up all the diverse choices that Drexler anticipates, challenge the *ability* of human associations to perpetuate themselves without altering the fundamental *need* for associations to do so, given the existence of competing groups with different ideas of what should be done. As a result, Stephenson must go deeper than Drexler into the question of how the "inward" side of nano-constructed lives, be they understood as the cultures behind the active shields or the lives of individuals within them, mold and are molded by the "outward" side: relations with other cultures and individuals.

We will explore various factors that create this situation, but competition and scarcity are key elements that Stephenson draws from Drexler's account. In a world of ongoing competition and still-limited (even if amazing) resources, how a given group answers the question of *what should be done* will be a factor in their success or failure vis-à-vis other groups with different answers. Stephenson also identifies factors other than scarcity and competition that mold the deployment and development of nanotechnology, and these round out Drexler's rather abstract picture of human things. In *The Diamond Age*, even when people try to make the kinds of fact-based rational decisions that Drexler would have us aim at, they still run up against the consequences of imperfect information and the misapprehensions it creates. And in Stephenson's fictional world, as in ours, people are powerfully moved by the love of their own, a love that eventually points to certain mysteries of the human heart.

OUR STORY THUS FAR

A very spare plot summary can only hint at the Trollopian, if not Dickensian, richness in character and incident of *The Diamond Age*. It is largely the story of Nell (short for Nellodee), an abused thete child, who at the age of five or six receives from her delinquent brother Harv (short for Harvard) a stolen copy of an extraordinary interactive, educational book/computer/game of immense power, *The Young Lady's Illustrated Primer*. Harv stole it from its designer, a talented nano-engineer named John Percival Hackworth, who intended it as a gift for his daughter Fiona. But Hackworth's copy is in turn a bootlegged version of the original, which he had designed for Alexander Chung-Sik Finkle-McGraw, one of the "duke-level Equity Lords" in New Atlantis.[94] Lord Finkle-McGraw intends it as a gift for his granddaughter Elizabeth, a gift he hopes will subvert her conventional New Atlantis education so that she can lead an "interesting" life; as one of the founders of New Atlantis, he finds his own children painfully dull and complacent.

Bootlegging the Primer requires that Hackworth use the facilities of Dr. X, a Mandarin working on three projects that move the plot: to free China from foreign influence, to rescue abandoned female babies from the ecologically collapsing Chinese interior, and to replace the centralized Feed of raw materials with "Seed" technology, decentralized nanotech manufacturing. Dr. X believes that the Seed will allow the Celestial Kingdom to become a true Confucian regime. Under Dr. X's power, Hackworth is enlisted more or less against his will to create the Seed.

The Primer, while an impressive example of artificial intelligence, still requires a human voice to interact with Nell as it tells her a host of compelling and didactic adventure-puzzle stories about a character who shares the name Nell.[95] The human voice is supplied by Miranda Redpath, a "ractor" (interactive actor) who comes to think of herself as educating and raising Nell, with whom she shares a background of abuse. (Even though Miranda has no idea who or even where Nell is, the Primer closely monitors and draws on the circumstances of Nell's real life, so Miranda can infer a great deal about what is happening to the real Nell from the lines she is given to read for stories involving the fictional Nell.) The net result of the Primer's education is that Nell turns into a formidable, self-reliant child who can escape her dead-end thete world when her life is threatened by one of her mother's abusive boyfriends. She has behind-the-scenes help from Finkle-McGraw, who pays the fees for Miranda's racting. He later supports Nell as well as Hackworth's daughter Fiona at Miss Matheson's Academy of the Three Graces, the posh neo-Victorian finishing school that his own granddaughter Elizabeth attends. While each of the three girls was educated by the Primer as the heroine of her own story, it is Nell who turns into the most impressive and unusual young lady. In fact, by the end of *The Diamond Age* she is a Queen of her own phyle, just like the character Nell in the Primer's didactic fairytale. Her real-life subjects are the abandoned girls saved by Dr. X. These hundreds of thousands of girls were also educated by pirated copies of the Primer that Dr. X and Hackworth supplied; these girls' experiences with the Primer have turned them into a real-world equivalent of the "Mouse Army" in Nell's Primer's story, an army that is entirely loyal to Nell.

While Hackworth's daughter Fiona and Nell are growing up, Hackworth spends ten years involuntarily (it would seem) immersed in the world of the Drummers, a strange underwater phyle whose members constitute a "Wet Net" in which individuals lose all self-consciousness in a quasi-telepathic linkage. Each becomes a node for processing information through intensive exchange of nanomachines in bodily fluids. Rescued once, Hackworth is drawn back to the Drummers for a second attempt to complete the Seed calculations in the midst of the Confucian uprising Dr. X has arranged. But while the last foreigners are expelled from China, many of their lives saved by Queen Nell and her real-world Mouse Army, Hackworth is again prevented from finishing the Seed calculations. For Miranda, having joined the Drummers in hopes that she could use their group mind to find Nell, was to be the sacrificial repository for this second attempt at Seed design. With the help of Carl Hollywood, Miranda's onetime producer, Nell locates and saves the woman whom the Primer itself has helped her understand as the loving human presence behind her book.

CONTINUING SCARCITY AND COMPETITIVE PRESSURES

Competition and conflict drive the plot of *The Diamond Age*, as China yet again seeks to get out from under foreign domination. Technology makes new forms of exploitation possible, but those new forms of exploitation mimic in their effects and results familiar stories from history. As Drexler seems to expect, individuals can find a social world into which they fit with least conflict by having choices with respect to phyle membership (if the desired phyle will have them, which is far from a foregone conclusion). However, as Drexler does not expect, the competing notions of a good life represented by the different phyles continue to be the basis for conflict among them in part because (as we will see) nanotechnology has not ended scarcity.

Competition among phyles has contrasting consequences; it usually, but not always, promotes Drexlerian diversity. In some instances it leads to an emulation of successful models and a certain homogenization. We learn, for instance, from Madame Ping, the proprietress of a brothel, that men "from all tribes" want "to be like Victorian gentlemen," so their sexual fantasies turn toward neo-Victorian scenarios.[96]

Competition among phyles means that they cannot in their internal organizations be blind to the qualities that will allow them to compete successfully. Observing the diversity of ways of life in her world, the teacher Miss Matheson notes as if speaking to Drexler himself: "It is upon moral qualities that a society is ultimately founded. All the prosperity and technological sophistication in the world is of no use without that foundation—we learned that in the late twentieth century, when it became unfashionable to teach these things."[97] Elsewhere in the book we learn more of this history: the New Atlantis phyle arose as a reaction against the moral relativism and mindless egalitarianism of the late twentieth century, just as the original Victorians turned against the excesses of the Regency era. Lord Finkle-McGraw grew up in Iowa (more or less in our present) and as a young man he

> had some measure of the infuriating trait that causes a young man to be a nonconformist for its own sake and found that the surest way to shock most people, in those days, was to believe that some kinds of behavior were bad and others good, and that it was reasonable to live one's life accordingly. . . . Finkle-McGraw began to develop an opinion that was to shape his political views in later years, namely, that while people were not *genetically* different, they were *culturally* as different as they could possibly be,[98] and that some cultures were simply better than others. This was not a subjective value judgment, merely an observation that some cultures thrived and expanded while others failed. It was a view implicitly shared by nearly everyone but, in those days, never voiced.[99]

The fashionable egalitarian relativism out of which New Atlantis sprang amounts to the belief (implicit also in Drexler) that one way of life is as good as any other. In the world of *The Diamond Age*, thetes continue to act on the basis of this mistaken belief, and we can see its ruinous effects. Nell's mother breaks up with her most promising boyfriend because he is a blacksmith in a tribe of artisans; they produce handmade goods for the neo-Victorians. She "didn't like craftsmen, she said, because they were too much like actual Victorians, always spouting all kinds of crap about how one thing was better than another thing, which eventually led, she explained, to the belief that some people were better than others."[100] From all we see of the thetes in the Leased Territories, we have no reason to believe her views are anything but typical. Certainly Nell's life as a thete, governed by violence, and the life of her thuggish father Bud, whom we see at the beginning of the book, represent a broader consequence of holding this belief, which is exploitation of the weak by the strong. The "logical conclusion" of the thete lifestyle is to be "homeless, addicted, hounded by debtors, or on the run from the law or abusive members of their own families."[101]

So despite nanotechnology's potential to greatly reduce material scarcity—no one need go hungry ever again, since public matter compilers make such things as food and first-aid supplies freely available[102]—inequality and relative poverty still exist.[103] Yet the failure of the lives lived on thete assumptions is not entirely a product of their own moral failings; it is also a consequence of those failings in connection with their external circumstances, over which they have little control. The lowland territories where the thete underclass lives provide a useful buffer zone against nano attack for the highland-dwelling Atlantans, even at the cost of chronic lung disease for its residents.[104] The thetes inadvertently are a sort of human layer in an active shield, a result of New Atlantis protecting its own interests. Nanotechnology is a big, powerful business that makes New Atlantis big and powerful. The Feed infrastructure was developed by and is owned and run by people made wealthy and powerful by its success; New Chusan is in effect Finkle-McGraw's ducal estate.[105] The interests that brought about this new world do not simply vanish when it is created, but continue to shape it. New Atlantis, in short, has no interest in giving all possible goods away for free; to some extent the Atlantans must manufacture scarcity along with everything else.

LOVE OF ONE'S OWN

It should not be news that people tend to protect their own interests; it is a manifestation of the love of one's own that is pervasive in *The Diamond Age*. Hackworth goes outside the law to give his daughter a leg up, yet repeatedly also expresses loyalty to his phyle; Harv severely beats and perhaps kills his mother's boyfriend, who has abused Nell; Finkle-McGraw wants the Primer to improve the life of his granddaughter and advance the interests of his phyle.[106] Carl Hollywood looks out for Miranda, his employee and perhaps romantic interest; Miranda comes to think of Nell as her daughter; the Mouse Army will do anything for their Queen Nell.[107] As these examples suggest, while love of one's own forms a basis for phyles that protects them from outsiders, it can also threaten their internal organization. This perennial tension inherent in voluntary human associations is not changed by technological development. Hackworth is loyal to his phyle in relation to outside phyles; he is loyal to his family inside the phyle; his wife might claim he is loyal to his own love of engineering problem-solving, putting his family on the outside. It is all the more a challenge to reconcile inner and outer when the definitions of each can shift in this way. The multiple layers of the love of one's own make relevant a variety of moral qualities that contain it or direct its expression, virtues like loyalty, honor, courage. Having been well instructed throughout the book in how phyles take care of their own, we are hardly surprised when a grandmother blows herself up to save her fleeing cohort from oncoming Celestial Kingdom troops.[108]

Sometimes the love of one's own directs violence against outsiders, sometimes it contains it in relation to those inside the group. It may involve competition among phyles, but there is also competition within phyles that must be dealt with. People in whom the moral qualities that discipline, train, and direct love of one's own are missing become dangerous; here again the thetes are instructive. Nell's mother and father have very limited attachments to anything beyond themselves. One of the reasons Bud is sentenced to death for a mugging is that he is not aware that his girlfriend Tequila has given birth to Nell; hence, the judge is confident that Bud's departure will not be a loss to his children.[109] While young neo-Victorian ladies are taught to think of late-twentieth-century urban America as a historical low point,[110] the thete lives we see are pretty nasty and Bud's is also literally poor, brutish, solitary, and short. Given their nearly unmediated selfishness, being spared the perils of starvation and other kinds of gross material deprivation by the free goods offered by the matter compilers only seems to liberate thetes to be more self-indulgent, irresponsible, and depraved.

But if thetes have little concern for anything outside of their own gratification, the other extreme is represented by the Drummers. In this tribe, individuals become parts in a "gestalt society," which is to say they share thoughts without being aware they are doing so.[111] By their appearance, behavior, and tunnel-dwelling lifestyle, the Drummers rather remind one of naked mole rats. So far as an outsider can tell they lack any sign of inner life, consciousness, self-consciousness or individual volition; it is hard to see how there would be any competition among them. (The few we see escaping the Drummers seem to do so only with outside assistance.) The mechanisms of their physical survival in this state—like what and how they eat—are largely left unexplained.

The ultimate purpose of their common life is similarly mysterious. Carl Hollywood concludes that by directly linking computing technology with the human mind, which "didn't work like a digital computer and was capable of doing some funny things," they have gained the capacity to break the encryption codes upon which the security of the Net depends—a development with disastrous potential for the status quo.[112] They have no "obvious way" or no desire to take advantage of their ability.[113] Yet we know they *are* willing to be used to create Seed technology, a tool equally subversive of the status quo. In losing a self-conscious identity, a core sense of being individuals, it seems the Drummers have succeeded either in transforming themselves into tools, the means to someone else's ends, or into an organization whose purposes are not comprehensible to those human beings who have not taken the Drummers' evolutionary leap to inhumanity.[114]

In different senses, then, both extreme love of one's own (as in the case of the thetes) and extreme abandonment of it (as in the case of the Drummers) are dehumanizing. It is the middle ground that brings us to the question of *perpetuation*: how a regime strives to contain and employ love of one's own to continue its institutions, fundamental ideas, and mores. For Miss Matheson, managing the tensions inherent in the love of one's own is a key element of perpetuation. Her work of "propagation" in her school is predicated on the belief that since it is a dangerous world, special care must be taken that, for "educated Westerners," the love of their own be properly constrained and directed to the group by what amounts to propaganda disguised as education and by inculcating the intensive discipline necessary to adhere to the powerful social norms of the neo-Victorians.[115] While Miss Matheson appreciates that for some outstanding individuals (such as Nell) more is possible, she would seem to be content if most of her students turn out to be relatively uninteresting because of their thoughtless acceptance of the norms of their world. She wants them to love New Atlantis because it is theirs.

Lord Finkle-McGraw, on the other hand, hopes the Primer will promote an education subversive of the complacency inherent in Miss Matheson's method, so that those who get it will come to love New Atlantis because they understand it is "the best of all possible tribes."[116] (Miss Matheson, in contrast, thinks it is only "as good as any other—better than most, really."[117]) Both Miss Matheson and Finkle-McGraw are dealing with the problem of perpetuation for a small portion of the ruling class of New Atlantis. Finkle-McGraw thinks that, for some part of this class, perpetuation is best achieved by a subversive education that will allow those who get it to maintain a sufficiently critical distance from their society—based on having the capabilities to live an "interesting life"—to allow them to choose it as best. For Miss Matheson, perpetuation is best achieved for most of this class by discipline and indoctrination. The ideas of both are consistent with the broad character of their New Atlantis phyle. Miss Matheson's school prepares its female charges for the phyle's complex internal norms of self-discipline and social interaction, ruthlessly enforced by peer pressure. Finkle-McGraw is evidently looking to what is required when commercial and technological competition and innovation are crucial to keeping the phyle on top. Their positions could be reconciled, of course, by discrimination between those who would be best served by each kind of upbringing.

Dealing properly with love of one's own is necessarily a difficult proposition. Thetes seemingly solve the problem by caring for nothing beyond themselves, with the consequent dangerous atomism. Drummers seemingly solve the problem by becoming an aggregation whose parts are unaware of having inner or outer lives. Stephenson acknowledges that nano and net technology could combine to liberate love of one's own entirely or destroy it, but that these are both dehumanizing choices. Between them rests the challenge of perpetuation, where moral qualities make the difference between success or failure at the difficult task of reconciling inner social and personal character with the outward circumstances of a competitive world.

THE INERTIA OF BEING HUMAN

The Diamond Age presents a world that is exactly like our world in that what is going on in one culture may not be comprehensible when viewed from the outside by another. A good deal of misunderstanding can be generated on this basis. But the same is true for one human being observing another; for all its technology, in Stephenson's world some mystery remains in the human heart. At one point in the story, Miranda is told by a certain Mr. Beck that it may be possible for her to locate Nell despite all the security built into the Net because that security is based on physical laws. But perhaps, he says, there is another dimension "invisible to those laws of physics, describing the same things with different rules, and those rules are written in our hearts, in a deep place where we cannot go and read them except in our dreams."[118] Beck, we find out, is a CryptNet agent, probably recruiting for the Drummers, who will use Miranda, perhaps precisely because of her connection to Nell and through Nell to Hackworth—or at least so we might speculate.[119] His appeal may not be disinterested, but it could still be true; after all, in the end Nell and Miranda *are* united.

Yet while the Drummers are able, it seems, to manipulate dreams and perhaps therefore hearts, they are not so attuned to the mysteries thereof that they can forestall the rescue of Miranda.

Mysteries of the heart come together with competition and love of one's own in a second shadowy conflict that moves much of the action of the book: the battle between New Atlantis and CryptNet. The New Atlantan view that there is a legitimate "cultural role in deciding what *should* be done" is challenged by the members of the CryptNet phyle, who believe, as Hackworth explains, "that information has an almost mystical power of free flow and self-replication, as water seeks its own level or sparks fly upward—and lacking any moral code, they confuse inevitability with Right."[120] Like Drexler, and reflecting many of the advocates of the eclipse of man, CryptNet apparently feels that changes brought by technology are both inevitable and the basis for "a more highly evolved society."[121] But, at least in Stephenson's vision, technological development does not so much lead to "higher" social forms as it shifts the shape of the constellation in which perennial characteristics of human life show themselves. Hackworth's own metaphor reminds us of the perennial wisdom in the book of Job: "man is born unto trouble, as the sparks fly upward."[122]

The CryptNet view—the idea that technological advancement equals inevitable social improvement—fails to appreciate what seems to be a key lesson of *The Diamond Age*: the complexly mixed human motivations that produce a given world-altering technological development do not vanish as that development works its changes upon the world. Born of desires for independence and control, of curiosity and mastery, the Seed technology that China and CryptNet desire for essentially contradictory reasons will not supersede or nullify such motives. Rather, its use will be conditioned by them, and by the myriad other sources of human action that come into play as a result of its employment by real people.

HUMAN NATURE, RESPONSIBILITY, AND TECHNOLOGY

The Diamond Age illustrates how the deeper and more nuanced one's understanding of human things, the more it makes sense to imagine technological change within the framework of historical or cultural continuity, and the less the discontinuities promised by the eclipse of man seem desirable. Stephenson does not deny that technology could usher in vast changes in the ways in which people live, even changes which may fundamentally alter our humanity. The kind of world depicted in *The Diamond Age* could represent an unstable, transitory, or intermediate phase of developments that will work themselves out in the victory of more radical possibilities. That outcome seems to be acknowledged in the book by the fact that none of the fundamental issues that move the plot is resolved. China, for example, is freed of foreigners, but its independence is by no means assured since it faces accelerated ecological collapse in its interior without either Feed or Seed.

Nevertheless, Stephenson imagines more convincingly than Drexler how deliberate choices with respect to what ought to be done with technology remain possible, and indeed how, given the enduring characteristics of human life, the increased power given by new technology will *require* greater care with making those choices if societies are to thrive. Stephenson goes beyond Drexler's acknowledgment of human failings to think about a future where a full measure of human motivations, noble and ignoble, continue to exist. Our frailties are not problems to be solved but are built into what we are and how we are placed in relation to each other, guaranteeing that perpetuation will remain a challenge. *The Diamond Age* begins and ends with the sounds of the bells of St. Mark's Cathedral in New Chusan—perhaps a reminder that godlike powers will not make us gods, let alone God.

Drexler's abstract understanding of humanity as "pattern plus stuff" in *Engines of Creation*—an understanding that makes our bodies, for example, somehow accidental to who we are—makes it easy for him to downplay the significance of the complex passions and interests out of which any real world of nanotechnology is likely to be built, even though he is not blind to them. The net result is his belief that nanotechnology will not only solve many of the problems of our world today, but that nanotechnology will solve nearly all the problems created by nanotechnology. In and of itself that prospect seems very unlikely—unless, as Drexler only suggests, it would result from competition among various ways of life effectively ceasing when human power reaches the limits of the possible. But in fact, so long as humans remain flawed in ways Drexler acknowledges, and in others he does not, as the realm of choice widens we will only have more reason to think about responsible choice—choice made with an eye to the technical facts, but also all the messy complications of individual and collective moral life.

Drexler is no Bernal or Haldane; his advocacy of dehumanization is considerably less overt. But there are surely more than hints of it in disembodied brains, in the bizarre experiments that nanotechnology will make possible, in mind emerging in machines, in the promise of remaking our bodies, and in all the essential premises of his argument. Stephenson seems to be suggesting that the inertia of human things will and ought to be a challenge to most of the radical achievements Drexler looks forward to. One could accept that analysis, and conclude: so much the worse for human things if they stand in the way of Drexlerian imagination.

With that thought in mind, we turn next to a more detailed examination of the transhumanists, to see what the eclipse of man has come to look like today.

CHAPTER FOUR

Perfecting Inhumanity

PROLOGUE: IT'S A WONDERFUL LIFE

At the age of 75, Adam Newman is deeply satisfied that with the onset of middle age he is reaching the full height of his operatic powers. He has always been thankful for the genetic endowment his parents arranged for him: the propensities for musical abilities and language skills, of course, but also the fact that they had enough foresight to keep him a bit over the average in height. His 6′6″ frame can carry with ease the expanded lung capacity and musculature that have made him such a powerful singer, and being big along with handsome still has substantial social and stage advantages.

Not that Adam has failed to make some improvements of his own. It was no picnic training to use his new optimized vocal cords, but well worth it in the end; many audiences still appreciate that it is really him they are hearing across his full six-octave range, and not some processed sound coming out of speakers. Of course the biomechanical cords also allow him to play the more traditionally minded houses that won't synthesize at all. His language and vocal-style memory implants are another matter; here again he can hardly fail but be grateful to his parents. For his innate skills in these areas mean that unlike many of his peers, who need a minute or two to reorient, he can "switch gears" seamlessly from one language or musical style to another, no matter how exotic. Since stylistic and linguistic eclecticism have run riot among composers (why not say what you want to say in whatever language that it sings best in?), this ability puts him in great demand. He does not need the contrived off-stage moments that librettists often have to supply for singers to recover after they have switched over. Adam feels it just adds to the realism.

He's never happier than when he is on stage; indeed, Adam can hardly imagine a different life for himself. Still, even in the very enjoyment of full summer Adam is sometimes reminded that fall will come. Lately, he's found that he really benefits from a couple of hours of sleep a week. (His folks once told him they had to stretch a bit to afford the sleep-optional mod, but thought in the end it was worth it—what great people they were!) And his metabolism has needed a bit of tweaking to optimize his nanotech waste-minimization feature. But, thank Gentech, so far it's nothing that remotely affects his voice. Adam has reached the point where he has no real concerns about his legacy; after all, his great performances are already available in full-def 3D VR, and he is under contract for a series that includes the new emotional track, having finally found a producer who shares his contempt for the crude work of much that is done in this area and is willing to listen to his suggestions about whom to use as his emo-dubber. (It never occurred to Adam that anybody would want to feel what he is *actually* feeling as he sings, which to be honest is not very much beyond a desire to please the audience.)

But he's beginning to wonder why he should have to think in terms of a "legacy" at all. Adam has never been what you would call an early adopter. He has all the innate conservatism you'd expect from an opera singer, but he has watched with interest the developing debate between the cyborgs and virtuals—those who have left birthbodies behind and uploaded their minds to biomechanical bodies and those who have chosen virtual instantiations. They both make good cases for their choices, although when the virtuals were just holograms, cyborgs seemed to have the better life. But then smartdust came along. Intellectually he knew Daphne Morgan wasn't a hologram anymore, but he never got over his surprise at how solid she was in that final embrace in *The Golden Ass*, when just a moment before she had been a tree. No more need to mime body-resistance with her! Still, the virtuals tend to get a bit . . . weird. Adam had been content with their collegial relation, but then she started giving him what he thought were signals that she wanted more. So he invited Daphne out to dinner to test the waters. He supposes he should be grateful that she came as a baby rather than adult giraffe. But having her standing across the table from him, with her big eyes, long lashes, and baby-doll lisp; it aroused in him feelings he did not quite know what to do with. Although it was supposedly impolite to ask them, everybody knew that the virtuals are multitasking their avatars when they are dealing with the very slow-paced solid world. The rumors about how virtuals like to mess with actuals, and about the strange things that virtuals get up to in cyberspace didn't help matters; he'd had a sense a couple of times that they were not exactly alone at the table. After that, he was content that they just work together.

So Adam wants to see how things shake out before he gives up entirely on the body that has been so good to him—just another way of showing gratitude to his wonderful parents. He wouldn't be a bit surprised if by the time he's ready, you won't even have to choose between virtual and cyborg; there will be something better than both! What would it be like to sing with a full cast of Adam Newmans?

DEHUMANIZATION IS central to contemporary transhumanism. A skeptical reader of this book's first chapter might have concluded that the advocates of the eclipse of man discussed there—Bernal, Haldane, and the rest—represent little more than historical curiosities. But if their ideas remain in some way curious, at least they are not simply historical curiosities. As we saw in Chapters Two and Three, the themes they explored are alive and well today: scarcity and competition, remaking our bodies and extending our lives, merging minds and machines, mankind's destiny in space, and more. These are some of the dearest hopes of today's transhumanists. Ray Kurzweil, the most prominent popularizer of transhumanist ideas, imagines we will have multiple virtual avatars, so that "I" can at the same time be attending various meetings and sharing a stimulating simulated sexual encounter that will be better than the real thing because I will be able to feel not only my own sensations and emotions but also those of my partner(s).[1] Hans Moravec, a prominent roboticist at Carnegie Mellon University, has enthused about how we will be able to speed up our thought processes so that "You will have time to read and ponder an entire on-line etiquette book when you find yourself in an awkward social situation."[2] (Of course, that seems to assume that one's fellows are not similarly "speeded up.") The transhumanist Simon Young looks forward to being able to eliminate the need to eat or excrete or sleep.[3] James Hughes, another booster of transhumanism, thinks we will have cybernetic brain implants that we will "undoubtedly" use for "watching and advising our behavior," to make sure we are acting morally by whatever standard of morality one might choose.[4] A 2006 conference at Stanford University Law School organized by some of the top figures in transhumanism featured a panel discussion about people like Cat Man and Lizard Man, who have had themselves surgically and otherwise altered to resemble their namesakes.[5] The list of alterations to our bodies and brains, our minds and morals, that the transhumanists envision goes on and on, becoming ever more strange.

Yet the best case for transhumanism starts from familiar, even prosaic, premises. This "common sense" case may be characterized as follows. Incremental technological and scientific progress, likely accelerating in pace, will continue to occur across a wide variety of fronts, driven by commercial, medical, and military motives, as well as the pure joy of research and development. Who does not already have a sense that things are changing rapidly, and that the rate of change is increasing? The consequences are readily taken for granted today: ever more people living lives that are healthier, wealthier, and longer. Who doubts that medicine will continue to make the kinds of strides that will offer opportunities for ever better health? Who would find it remotely plausible that computers would be the same in a decade as they are today?

Hence, many of the new abilities made possible by this progress will, taken one by one, hardly be controversial in themselves.[6] But taken together, these ever-increasing powers over nature will converge on the ability to redesign our minds and bodies, opening the door to the more strange and radical possibilities mentioned above. There need be no dedicated research program to make these dreams a reality; the necessary technologies will appear on their own, regardless of transhumanists' hopes for the future.[7] The technologies that cure disease and restore lost functionality open the door to enhancement of our abilities; indeed it seems to be agreed by both transhumanists and many of their critics that there is no bright line between therapy and enhancement.[8] So transhumanists readily imagine using our increasingly sophisticated understanding of the human machine to create novel capacities.

Indeed, the more one thinks about how much better we could do if we designed our own bodies, the more dissatisfied we are likely to be with the present model. In particular, isn't just about everybody dissatisfied with the brevity of life? Pushing this open door is a favorite transhumanist selling point; they are confident that significant life extension is on its way. Ongoing developments in our ability to treat once-fatal diseases will certainly play a role, but aging itself, seen as a disease, will soon enough start to yield, whether to genetic engineering, artificial or cloned organs, or the kind of nanotechnology-enabled medicine that Drexler foresaw.

As we will see in what follows, the arguments that drive transhumanism from the uncontroversial to the strange are very similar to those that moved us from Condorcet's vision of human progress to the eclipse of man. Transhumanists argue, in a fashion that will by now seem familiar to readers of this book, that manipulating nature is simply part of what defines us as human beings. The growing abilities of modern science and technology that stem from this trait are giving us the power to take evolution into our own hands; ongoing competition will force us to use technological evolution to improve not only on the naturally given "outside" of us but also on our selves.

EAGERLY AWAITING ENHANCEMENTS

Let us begin by looking more closely at the first part of this line of thought. There are already many technologies available that enhance human abilities. Take, for example, human vision. Eyeglasses enhance the vision of those with poor eyesight, but they are already coming to appear as relatively primitive. Contact lenses can both enhance eyesight and provide cosmetic alteration of eye color or appearance. LASIK surgery can eliminate the need for certain kinds of vision prosthetics altogether. Artificial retinas are being used to restore sight to people with certain kinds of blindness, and tiny "telescopes" are being implanted into the eyes of some patients with macular degeneration.[9] Such technologies are rapidly improving, and remarkable further advances are both easy to imagine and plausible. None of this "enhancement," if that is indeed what it is, is remotely controversial, and we expect and desire further improvements in such technology, made possible by growing knowledge of how vision and the eye work, and by advances in computing, materials science, and biotechnology. So an artificial eye, whether grown or constructed with organic or inorganic materials, seems far from a crazy dream, but rather a reasonable goal of medical research and a great boon, restoring normal sight to all those who suffer from vision loss.

So far so good. But if one can make such an eye, why restrict its use to those with vision loss? Just as many cars now come equipped with once-unimaginable rear-vision cameras, would it not be useful to implant a third eye in the back of one's head? Granted, the brain is not used to processing the data such an eye would provide. But the brain is quite malleable; could it not be trained to see with a third eye? A recent study suggests it treats tools as body parts, so why not new body parts as tools?[10]

If there are atavistic aesthetic objections to a third eye, why not construct an eye that has *better*-than-normal vision, one that can see more sharply at greater distances, or can see better in low light?[11] Why not an eye with a zoom lens, or one that can extend the range of human vision into the ultraviolet or infrared? We are told that the eye is a kind of camera, and we already make cameras that do all these things. The brain "processes" electrical impulses from the eye, and we now also make cameras whose outputs are electrical impulses.

It does not take much imagination to see how such artificial, enhanced eyes could provide a competitive advantage to soldiers, pilots, police, sportsmen, contractors, repairmen—the list could be as long as the kinds of additional vision that *someone* might find useful. And once some people adopt the new ability, others in their line of work will be faced with competitive pressure to keep up by using the same enhancement or risk falling behind. While doubtless early versions would be quite expensive, why not experiment on wealthy early adopters? Just as the price of LASIK surgery has plummeted, so too we can expect that some kinds of artificial eyes will eventually come to be affordable by nearly all—even if the cutting-edge models continue to be high-priced.[12]

The transhumanists can point out that all you have to do is look around you today to see that if you leave people free to make their own choices about what kind of vision or any other sort of enhancement they wish to have, things will settle out in a way that favors their ever-increasing use, even if the distribution of benefits remains inegalitarian.[13] This argument, of course, tacitly acknowledges both that there will be winners and losers in a transhumanist future, and who the losers will be: the unenhanced or even the less-enhanced.[14] Should we consider that a problem? Many (though by no means all) people accept that there will be winners and losers in an economy based on just the kind of free choice that the transhumanists advocate in relation to enhancements. So it seems as if anyone already accepting such inequalities would have to do so also in relationship to enhancements.

But there is a difference. The present inequalities are generally accepted to the extent that we believe winners and losers are not graven in stone, that there is dynamism and mobility such that winners are not entirely self-perpetuating—for example, there are reversals of fortune, or successful people can have stupid and profligate children who fritter away the fruits of their success. Beyond that, mortality is a great leveler. But transhumanist success aims precisely to be self-perpetuating to a far greater degree and takes perpetuity of some sort to be a reasonable aspiration. Hence, the prospects for entrenched inequalities seem to be dramatically improved as the logic behind how an artificial eye might be used is applied to just about any aspect of our bodies: senses, limbs, or organs, including our brains.

CHOOSE YOUR OWN ADVENTURE

Nevertheless, in relation to all such technologies of enhancement, a great many transhumanists stand foursquare behind the principle of consumer choice. Most are willing to concede that enhancements ought to be demonstrably safe and effective. But the core belief is that people ought to be able to choose for themselves the manner in which they enhance or modify their own bodies. If we are to use technology to be the best we can be, each of us must be free to decide for himself what "best" means and nobody should be able to stop us.[15]

This techno-libertarianism is important to transhumanists if for no other reason than that it allows them to believe they can distinguish between their outlook and that of the early-twentieth-century advocates of eugenics with whom they fear being confused. After all, the transhumanists, like the eugenicists before them, talk about a quasi-Darwinian competitive imperative to improve the human race; it might appear as if transhumanists are simply advocating more effective means to old-fashioned eugenic ends, replacing selective breeding and sterilization with enhancements. But transhumanists insist it is crucial that, unlike the eugenicists, they are not interested in using the power of government to coerce people into making better humans. For the most part the transhumanists insist that just as no one should be prevented from choosing the enhancements or modifications he wishes, so no one should be forced into any kind of enhancement or modification.[16] Indeed, the transhumanists argue, it is their critics—whom they disparagingly label "bioconservatives" and "bioluddites"—who, by wishing to restrict enhancement choices, are the real heirs of the eugenicists; they are the ones who have an idea of what humans should be and want government to enforce it. The transhumanists would say that they are far less interested in asserting what human beings should *be* than in encouraging diverse exploration into what we might *become*, including of course not being human at all. Moreover, the argument goes, transhumanists are strictly speaking not like eugenicists because they are not interested only in making better human beings—not even supermen, really. For to be merely human is by definition to be defective.[17] It is this view of human things that makes the transhumanists *de facto* advocates of human extinction. Their dissatisfaction with the merely human is so great that they can barely bring themselves to imagine why anyone would make a rational decision to remain an unenhanced human, or human at all, once given a choice.[18]

However, if the transhumanists are for the most part against state coercion in relation to enhancements, as we have already seen that does not mean there is no coercive element in the transition to the transhuman. They can avoid government coercion because they believe that the freedom of some individuals to enhance and redesign as they please adds up to an aggregate necessity for human enhancement, given competitive pressure and the changing social norms it will bring. Indeed, to the extent that transhumanists recognize that theirs is presently the aspiration of a minority, they are *counting* on this kind of pressure to bring about the changes in attitude they desire. Within the framework of the largely free market in enhancements the transhumanists imagine, an arms-race logic will drive ever-newer enhancements, because if "we" don't do it first, "they" will, and then "we" will be in trouble. This kind of coercion is not of much concern to transhumanists; they are content to offer that it does not infringe upon freedom because, as the rules of the game change, one always retains the freedom to drop out. Indeed, the transhumanists seem to take particular delight in pointing out that anyone who opposes the idea that the indefinite extension of human life is a good thing will be perfectly free to die. In a world of enhancement competition, consistent "bioluddites" will be self-eliminating.[19]

There is another kind of necessity that transhumanists can adduce to support their case for the free choice of enhancements. Recalling a theme introduced by Condorcet, they assert that human beings are practically by definition the beings that enhance themselves, or that human beings are the beings who overcome their own limitations and the limits of the naturally given—including the sorts of limits that Malthus had in mind. According to British bioethicist John Harris, who for no good reason objects to being called a transhumanist,

> It is doubtful that there was ever a time in which we ape-descended persons were not striving for enhancement, trying to do things better and to better ourselves. . . . I am personally pleased that our ape ancestor lacked either the power or the imagination, or indeed avoided the errors of logic and/or morality, which might have led her to preserve herself at our expense. I hope that we will have the imagination, the power, and the courage to do better for ourselves and our descendants than the combination of chance, genes, and environment has done for us.[20]

While Harris frames the issue as a choice, it is not clear why we are any more free to make it than our ape ancestress, given his initial assumption about what it means to be human—that we have always been striving for enhancement. Starting from this definition, the transhumanists are simply doing what human beings have always done. So just as history tells the story of cultures gradually gaining the wisdom to accept as humans people who come in all kinds of shapes, sizes, abilities, and colors, we ought to be prepared to extend that circle of identification and accept further such variance so long as the essential attribute of self-overcoming is still present. Transhumanist philosopher Nick Bostrom has suggested that even if they are not human, transhumans and posthumans may yet be "humane"—that is, they may possess recognizably admirable human moral qualities. Indeed, they may well be more likely than humans to be humane, since they will have overcome the human or natural constraints that produce bad behavior among humans.[21]

Yet by calling attention to this self-overcoming aspect of human nature, the transhumanists create another kind of problem within their argument. Starting from their assumptions, evolution is the only way in which human beings could have gained their ability to overcome natural limits; that we have this capacity is ultimately a matter of chance. The ability to control our own evolution going forward, then, has at least this much of an irreducible foundation in natural evolution. That foundation would seem to represent a limit on what parts of our nature we would seek to modify, lest we compromise this all-too-human drive to self-overcoming. Or, to describe this paradox in another way, our descendants would only genuinely be "posthuman" if they reached the point where they were no longer engaged in the self-overcoming that is our presumed essence. Whether such a result would ever be desirable is unclear, unless lurking somewhere in the transhumanist argument, as it did in Drexler's, there is a tacit belief that at some future moment our descendants will have reached some limit of the possible with respect to what can be done to themselves. Only at that point would this drive for enhancement become counterproductive and have to be rooted out at last. Short of this outcome, a remnant of our nature remains in whatever kind of beings come after us.

GUARANTEED RETURNS?

Competition combines with the challenging of Malthusian limits to give the transhumanists a very strong sense that they represent the direction of history. Ray Kurzweil has taken the argument one step further, giving it a cosmic twist (like Drexler and Reade before him). Kurzweil calls the force that drives the progress of inhumanity the "law of accelerating returns." As its name implies, this law encapsulates the idea of an accelerating rate of progress, but Kurzweil claims that he can show the trend line is an asymptotic curve, rapidly accelerating today toward an unreachable future infinite. Something like this law, he seems to believe, has been at work extending even into the deep past, long before the rise of humanity.[22] If Kurzweil is correct, then any appearance of choice we might have with respect to a posthuman future is illusory, and the might of the law behind this trend line is the ultimate source of the rightness of the future he describes. (So Kurzweil, like CryptNet in *The Diamond Age*, confuses "inevitability with Right.") Why bother advocating for transhumanism at all, if there is no reason to fear that transhumanism's critics might win out, as the law dictates their irrelevance? Advocacy for transhumanism must instead be predicated on concern that without a sufficiently solid understanding of what the future holds, the powers we are gaining could be misdirected in destructive ways. In other words, instead of deliberate human extinction, we might cause it accidentally.

Still, perhaps the transhuman future need not be quite so dire. David Pearce, a cofounder of the World Transhumanist Association, has what at first glance seems to be a considerably more benign outlook. He returns us to the prosaic, seeing the transhumanist effort to control evolution as being driven by what he claims is a completely rational and undoubtedly moral desire to maximize happiness and minimize suffering and pain:

> Over the next thousand years or so, the biological substrates of suffering will be eradicated completely. "Physical" and "mental" pain alike are destined to disappear into evolutionary history. The biochemistry of everyday discontents will be genetically phased out too. Malaise will be replaced by the biochemistry of bliss. Matter and energy will be sculpted into life-loving super-beings animated by gradients of well-being. The states of mind of our descendants are likely to be incomprehensibly diverse by comparison with today. Yet all will share at least one common feature: a sublime and all-pervasive happiness.[23]

We would of course first make every effort to get our own house in order, making sure that human beings are all happy all the time, but there is also pain and suffering in other parts of the animal kingdom, and eventually we would face up to the moral necessity of dealing with that by the appropriate changes in genes and ecosystems to put an end to animal predation. Pearce goes on to consider what happens when we discover life on other worlds, and has no compunctions about extending his principle, and hence posthuman efforts, in that direction. "So long as sentient beings suffer extraordinary unpleasantness—whether on Earth or perhaps elsewhere—there is a presumptive case to eradicate such suffering wherever it is found."[24] When the posthuman Earth fleet arrives on an alien doorstep, it is only with the most benign and unselfish intention of completely restructuring their world so that all its beings can share in such happiness as we make possible for them. In Pearce's account of the posthuman future, our descendants become the aliens we have feared or hoped for. Yet if we imagine an alien fleet arriving on Earth with the same intention of rebuilding nature from the ground up, we are entitled to wonder if it would be taken as good news.

OLD WINE IN NEW BOTTLES

Meanwhile, the roboticist Hans Moravec has a different take on the transhuman future. As an alternative to Bernal's idea of transplanting an organic brain into a mechanical body, but based on a similar view of the primacy of the brain, Moravec seems to have been the first to think out in a systematic way what would be involved in "uploading" a mind into a computer. As he explains it in his classic 1988 book *Mind Children*, Moravec's premise is the same idea of "pattern identity" that Drexler spoke of; who we are is fundamentally a matter of the processes going on in our brain, the electrochemical activity among our neurons, not the stuff itself; "the rest is mere jelly."[25] There is of course a huge number of neurons with an even larger quantity of interconnections among them, so this electrochemical activity is extremely complex—but not infinitely so. Sooner or later, Moravec argues, as computing power grows there will be some ultra-sophisticated version of today's brain scanners that will make it possible to scan neuron by neuron what is going on in the brain, and store and duplicate those results in some combination of software and hardware simulation.

Moravec assumes that at least at first, efforts at uploading would have to destroy the brain bit by bit in order to expose each neuron for scanning. He asks us to consider someone hooked up to a device that would read and replicate each neuron. Suppose that as each cellular layer is scanned, the subject could switch back and forth between the organic brain and the simulation; if the system works correctly he could not tell the difference, so that each layer could be destroyed and the next scanned. As the process continued, more and more of the subject's thinking would be going on in the computer, until finally it would completely replace his old brain.[26]

Moravec imagines that the brain thus computerized could, in the manner anticipated by Bernal, be combined with any sort of robotic body one wished. His personal favorite seems to have been the "bush robot," which is little more than a collection of arms and manipulators arranged in a fractal pattern, from the big to the very tiny. He described it thus:

> A bush robot would be a marvel of surrealism to behold. Despite its structural resemblance to many living things, it would be unlike anything yet seen on earth. Its great intelligence, superb coordination, astronomical speed, and enormous sensitivity to its environment would enable it to constantly do something surprising, at the same time maintaining a perpetual gracefulness. . . . A trillion-limbed device, with a brain to match, is an entirely different order of being. Add to this the ability to fragment into a cloud of coordinated tiny fliers, and the laws of physics will seem to melt in the face of intention and will. As with no magician that ever was, impossible things will simply *happen* around a robot bush.[27]

The bush robot was not the end of his imaginings. Computerized minds could communicate with and access each other with all the ease of networked hardware and software. They could be duplicated, or optimized to run in various configurations. Beyond that, just as one might merge electronic documents, why not merge one electronic mind pattern with another? Since any brain could be scanned, why shouldn't a human pattern be merged with an animal pattern?[28]

Moravec points out that this computerized mind could be effectively immortal (assuming sufficient provision for backups), which might seem like a very appealing prospect. But he recognizes that this immortality might not end up being precisely what someone might have originally hoped for. Competition and scarcity ("changing conditions," "external challenges") mean that the I that sought immortality will not actually get it:

> Immortality of the type I have just described is only a temporary defense against the wanton loss of knowledge and function that is the worst aspect of personal death. In the long run, our

> survival will require changes that are not of our own choosing. Parts of us will have to be discarded and replaced by new parts to keep in step with changing conditions and evolving competitors. . . . In time, each of us will be a completely changed being, shaped more by external challenges than by our own desires. Our present memories and interests, having lost their relevance, will at best end up in a dusty archive, perhaps to be consulted once in a long while by a historian. Personal death as we know it differs from this inevitability only in its relative abruptness. Viewed this way, personal immortality by mind transplant is a technique whose primary benefit is to temporarily coddle the sensibility and sentimentality of individual humans.[29]

Note how the desire for personal immortality is such a small thing from Moravec's point of view, a sentiment to be coddled, a kind of temporary error. Here is surely Olympian detachment! It is as if the gods in the full enjoyment of their immortality would wonder why mortals make such a big deal about death. And yet somehow this small thing drives us on to great things.[30]

Given the speed with which computing technology developed since Moravec first published his thoughts on uploading in 1988, it did not take long for the idea to be modified to conform to new ideas and capacities. The rise of virtual reality made some wonder why the uploaded mind would require a physical instantiation at all when it could occupy whatever kind of virtual realities it wished. For those who might want to insist on continuing to be able to interact in the "real world" but find robotic bodies too constraining, the theoretical nanotech "smart dust" provides an imagined solution, promising "real" virtual bodies that would have none of the messy inconveniences of actual living things—very much along the lines that Fedorov imagined. And there is the added benefit that were such bodies ever to be disrupted in some way, they could always simply be rebooted.[31]

NEW WINE IN NEW BOTTLES

In Moravec's presentation, transhumanist aspirations do not end with the (somewhat deceptive) promise of personal immortality. Once biologically based human patterns are instantiated in computers, they will as we have already noted become upgradeable in all kinds of ways. In particular, they will be able to think faster. Hence, just as a chess-playing computer can today consider more options in greater depth and shorter time than a human player, so will the formerly human patterns be able to think about and experience far more than a human being—and this without fatigue, without forgetfulness, without the distractions of biological needs. Here is surely a great boost to progress in any area to which these posthumans may wish to turn their attention. Problems will be modeled, solutions simulated, results refined in an interactive cycle of astonishing speed and complexity. Such intelligence, alone or in concert with advanced artificial intelligence, would constitute what is sometimes called "superintelligence" or "hyperintelligence." Its arrival, as Drexler suggested and as many others now believe, would be just the sort of thing that would rapidly push technological development to the limits of the possible. As Bostrom has speculated, "Superintelligence would be the last invention biological man would ever need to make, since, by definition, it would be much better at inventing than we are."[32]

What exactly happens to humanity at that moment is an important and to some degree divisive issue for transhumanism. The issue is framed by the concept of "the Singularity." While the very meaning of this term is contested, it is fair to think about it in the way it was first systematically presented by mathematician and science fiction author Vernor Vinge:

> When greater-than-human intelligence drives progress, that progress will be much more rapid. In fact, there seems no reason why progress itself would not involve the creation of still more intelligent entities—on a still-shorter time scale. The

> best analogy that I see is with the evolutionary past: Animals can adapt to problems and make inventions, but often no faster than natural selection can do its work—the world acts as its own simulator in the case of natural selection. We humans have the ability to internalize the world and conduct "what if's" in our heads; we can solve many problems thousands of times faster than natural selection. Now, by creating the means to execute those simulations at much higher speeds, we are entering a regime as radically different from our human past as we humans are from the lower animals.
>
> From the human point of view this change will be a throwing away of all the previous rules, perhaps in the blink of an eye, an exponential runaway beyond any hope of control. Developments that before were thought might only happen in "a million years" (if ever) will likely happen in the next century. . . . I think it's fair to call this event a singularity ("the Singularity" for the purposes of this paper). It is a point where our models must be discarded and a new reality rules. As we move closer and closer to this point, it will loom vaster and vaster over human affairs till the notion becomes commonplace. Yet when it finally happens it may still be a great surprise and a greater unknown.[33]

If this hyperintelligence, like human intelligence, is routinely dissatisfied with its own capacities, it will surely turn its attention to finding the matter and energy for the ever-increasing computer processing power it will presumably demand. As Moravec puts it, "Our speculation ends in a supercivilization . . . constantly improving and extending itself, spreading outward from the sun, converting nonlife into mind."[34] Matter and energy are not in short supply; what already exists just needs to be used to some purpose. Think of all the mass wasted in asteroids, moons, lifeless planets, or indeed in organic bodies of all sorts that do not need to live in biological form when they could just be simulated. Think of all the radiation that streams pointlessly into space from stars, but could be captured by enclosing them in gigantic spheres.[35] Just as, from the transhumanist point of view, the human body is sub-optimally designed, so too there may be a great deal of work to be done to reconstruct our own and other solar systems to make them happier homes for intelligence. Whatever hyperintelligence is in its own way is, like the human mind, simply a pattern of information, a pattern that can be replicated in a variety of forms. The deconstruction of our solar system to turn dead matter into mind, and any other solar system hyperintelligent posthumanity may be moved to encounter,[36] would not have to mean the loss of the information it represents.[37] It was J. D. Bernal who said that "the desirable form of the humanly-controlled universe" is "nothing more nor less than art"—but in the vision of today's transhumanists, that desirable form is not, as it turns out, actually controlled by humans.[38]

WHEN HYPERINTELLIGENCE COMES KNOCKING

Ray Kurzweil takes the argument about the capacities of hyperintelligent posthumans one step further. He wonders whether our understanding of the constraints imposed on us by the laws of nature may just be an artifact of our limited intellectual abilities. For the sake of "saturating" the universe with intelligence, which he (like Haldane) believes is "our ultimate fate,"[39] Kurzweil hopes that the speed of light may not after all prove to be a cosmic speed limit.[40] Along with Moravec he would like to believe that efforts will be made to overcome the entropy that will eventually cause the universe to "run down."[41]

These remarkable aspirations clearly connect back with issues we already looked at in connection with SETI. If Kurzweil is correct about the potential abilities of hyperintelligent posthumans, then would the same not also be true of the hyperintelligent form of some alien race? If so, then Fermi's question arises once again—*where are they?*—and we probably have to have recourse to the same kinds of arguments already noted about our decidedly limited ability to detect the existence of such godlike beings, particularly if they wish to remain undetected. Then again, even less-capable aliens bound by the speed of light would still have had plenty of time to populate the galaxy, or use its mass and energy for computational purposes in a way that we might detect. As we saw in Chapter Two, such arguments make Kurzweil and some other transhumanists skeptical that there is alien intelligence at all.[42]

And perhaps we should be happy about that. If we take posthuman hyperintelligence and the Singularity seriously, we are saying that the successors we are being asked to open the door to will be alien to us far beyond any change possible in natural evolution. Vinge from the start understood that one of the greatest unknowns would be how hyperintelligent beings would treat human beings. Would we be seen as revered progenitors or as inconvenient pests? Such a question is hardly new to science fiction, of course, but now there are no few transhumanists and AI proponents who think that the question of "friendly" or "moral" AI needs to be brought to the fore of serious, real-world efforts.[43] What "friendly" or safe or moral AI would look like is anybody's guess, but it is very hard to see why the net result would not look very much like *Childhood's End*, given the radical lack of a common framework of judgment between us and hyperintelligent life. Kurzweil likens it to the gap between us and bacteria, and yet no one claims seriously that we share a common moral universe with bacteria.[44] Hence the transformation of supposed hyperintelligent benevolence into perceived malevolence remains a live issue. It does not seem likely that these otherwise incomprehensible beings will be more moral than we are in terms that we could appreciate.

So if this kind of posthuman hyperintelligence were to arrive on our doorsteps tomorrow, it is hard to see how it would look different from a hostile alien invasion of the sort classically depicted in science fiction. Separated from them by their Singularity (forget about mere biological or cultural differences), we would be incapable of understanding their needs and motivations beyond noting that they seemed to have an insatiable appetite for transforming matter and energy into forms useful to them but inimical to us. The opportunity to be absorbed (assimilated?) into whatever mode of life they will have created for themselves, to have our patterns preserved, would seem to be about the most we might expect.

TRANSHUMANIST TENSIONS

There are genuine tensions between various aspects of the case for transhumanism that we have highlighted, not the least of which is that something that begins by promising human enhancement ends up more than reconciled to human extinction. More specifically: on the one hand, the motive force for transforming ourselves is a deep dissatisfaction with the merely human. On the other hand, this dissatisfaction, and the efforts at transformation it produces, are presented as quintessentially human. We are all to be free to make choices about enhancements, but radical inequalities may lead some to be freer than others. To raise our hopes, the transhumanists urge overcoming human limitations. To quiet our fears, they claim there could be continuities between us and the superior beings to come. Specific human characteristics are treated as if they are merely contingent with respect to all the alternative ways we can imagine intelligence might be embodied, but however embodied we are told there is no reason to think that intelligence could not still be humane. Yet even the continuity of the humane gives way to the absolute discontinuity of the Singularity.

These tensions mean there are reasons to have qualms about the transhuman and posthuman future despite the fact that it can seem to be derived from minimally controversial premises. The prosaic best case does not quite tell the whole story, but is more like the proverbial slowly warming pan of water in which the frog sits unaware of the boiling point to come. By thinking through the assumptions of the eclipse of man all the way through to the Singularity—to a point where it must be acknowledged that none of us knows or could even understand the full implications of the path we are set upon—today's transhumanists have clarified the full implication of those assumptions even more than the earlier thinkers we examined. But that does not mean they have thought them through completely.

In the prologue to this chapter, the character of Adam Newman has not been dehumanized—or at least, not obviously so. But his story exemplifies some problems to come, should transhumanist aspirations prove possible. We don't know that talents and dispositions can be engineered in the way implied in the vignette, but on the assumption that they can be, is Adam the freely choosing transhumanist ideal? He seems very much an instrument of his parents' wishes for him, far more than is possible for even the most controlling parents today. There are hints that he is unable really to question that legacy; as a result he is less of a free man and self-overcomer than he himself might believe. We may envy the ease with which he falls into a role that is perfect for him and the happiness it provides him, and still wonder whether this lack of tension accounts for his emotional flatness. We can wonder if his musicianship represents the triumph of technique, with Adam occupying something like the position of a racecar driver, in effect the business end of a highly tuned team of technicians who have made him what he is. Yet he probably has less understanding of what they do and how they do it than the driver does.

Then there is the character of Daphne Morgan. She is not so dehumanized that we cannot suspect her of being a trickster or a tease, or of having the sort of contempt for "solids" that bohemians have for bourgeois. But if her own self-regard makes her no more humane than any morally fallible human being, surely she has far greater power and even motivation to exercise her inhumanity, to sport with Adam as the old gods used to sport with humanity. Surely Adam's self-satisfied narcissism would make him a tempting target. Yet his sense that there is more going on with her than he is likely to be able to understand is probably correct. How would Adam understand what it is like to be in more than one place at a time, or to have all his consciousness be as editable, reversible, and replayable as the recordings he makes?

Looking at such distant prospects, which is after all encouraged by the transhumanists themselves, reveals important problems with their arguments. And yet it could be objected that even if the transhumanists are correct about what the future holds, any problems associated with it will not be *our* problems—or at least will be experienced only by the relative few reading these words today who will have made provision for their own reanimation. For most of us, as technology progresses incrementally, the issues are going to be posed in terms of far more prosaic-seeming technologies of human enhancement, technologies that may look attractive whether or not there is an obvious link to the transhumanist aspiration for self-design. Hence, while there is no lack of thoughtful explorations of more extreme transhuman futures in science fiction (some of which we have already discussed in our examination of the "benevolent" superintelligence in *Childhood's End*), it makes sense to explore instead a still speculative, but not that speculative, nearer-term development: the ability selectively to erase unhappy memories. Difficulties with this relatively modest kind of enhancement will alert us to difficulties inherent in the more radical alterations transhumanism proposes.

NO THANKS FOR THE MEMORIES

As imagined biotechnologies go, memory alteration seems not quite in the same league as things like the transhumanist hopes for direct links between brains and computers, or wireless connections between brains that would allow people to feel each other's feelings, or the ability to "upload" into a machine the entire pattern of our identity. But it is worth considering memory alteration closely, since memory is central to identity, and identity to happiness—a point highlighted in *Beyond Therapy*, a 2003 report of the President's Council on Bioethics.[45] So if we are to make ourselves better off by the use of emerging enhancement technologies, or if in accord with Pearce's hedonic imperative we are to be happy all the time, memory and memory alteration in some form will be at the core of our various pursuits of happiness.

Beyond Therapy focuses in a general way on the use of drugs to head off the impact of likely traumatic memories, or reducing their subsequent emotional consequences.[46] Grounded in present medical realities, the report does not speculate at any length on the consequences of developing less blunt approaches to adding or subtracting specific memories as we understand the neuro-chemistry of the brain with increasing precision. While it is always easy to understate the difficulty of such efforts, there is nothing in our present understanding of the brain that definitively shows that precise interventions in memory will be impossible.

The consequences of selective forgetting have been a literary theme since long before science become seriously interested in the topic; think of magic potions in tales ranging from Homer to Wagner, or the elaborate deception practiced on the title character in Ermanno Wolf-Ferrari's opera *Sly*. Here we will look at two stories about memory alteration. The first is *Dr. Heidenhoff's Process* by Edward Bellamy, who would some years after writing it become internationally famous for his utopian novel *Looking Backward*. The second is the movie *Eternal Sunshine of the Spotless Mind*, which provides a corrective to the generally positive position taken on memory alteration by Bellamy.

Bellamy's 1880 novella *Dr. Heidenhoff's Process* is hardly great literature, but it nevertheless contains a well-thought-out moral defense of selective memory obliteration. If the story is Victorian in its operatic sentimentality, its effort to think through the moral consequences of Dr. Heidenhoff's process for elimination of troublesome memories exhibits likewise a Victorian intellectual thoroughness.

The plot is simple enough. Henry Burr loves Madeline Brand unrequitedly. Just as it seems she might be returning his feelings, the flashier Harrison Cordis comes to town, and Madeline and Harrison begin a flirtation. As their relationship deepens, Henry leaves town to work in Boston, unable to bear the loss of Madeline. However, "there was a scandal" and Harrison "deserted her."[47] A fallen woman, Madeline flees to Boston. Henry finds her and offers marriage, but in her shame she keeps putting him off; she cannot ruin him as she has been ruined. One day, she seems to relent, and even allows Henry a not-so-chaste kiss prior to their parting, promising to settle the matter the next day.

Henry returns to find her terribly excited; she has seen a story in the newspaper about a technique invented by a certain Dr. Heidenhoff for the purpose of selective memory destruction. Madeline is anxious to have the memory of her shame destroyed so she can marry Henry with a clear conscience. Skeptical at first, Henry visits Heidenhoff and comes away convinced that Madeline should indeed try the process; "the thought of receiving his wife to his arms as fresh and virgin in heart and memory as when her girlish beauty first entranced him, was very sweet to his imagination."[48] The procedure works, the couple are engaged—but at the moment of seeing Madeline model her wedding gown, Henry wakens to find he has been dreaming. He almost immediately receives Madeline's suicide note, in which she repeats how she could not dream of ruining him, and gently chides him for his failure to perceive that the kiss she at last permitted him was one of farewell.

THE SPONGE OF OBLIVION

Dr. Heidenhoff is no shrinking violet when it comes to explaining and defending his invention. In the full enjoyment of what he himself regards as the powers of a god, Heidenhoff is only too happy to explain the medical basis for his process; his hopes for "amusing experiments" that will confound the "thicker-headed" and "foggy-minded," whose false moral assumptions will be exposed; and the "good tidings" that his invention brings the world.[49]

Indeed, Heidenhoff is so obnoxiously full of himself that it is tempting to think that Bellamy might be interested in exposing scientific hubris. Yet while that possibility cannot be entirely eliminated, it seems unlikely given the fact that the story is framed by two suicides stemming from their perpetrators' inability to forget shameful events. *Dr. Heidenhoff's Process* suggests that neither love nor Christianity nor conventional moral ideals are adequate to the problems created for us by the memory of the misdeeds we do or those done to us. Bellamy evidently had some hope that science could step into the breach.

Dr. Heidenhoff believes that memory has a physical basis in the brain[50]—hardly a controversial proposition today. He admits that, for the time being, his process, which applies electricity to the brain, can only extirpate memories that have been so dwelled upon as to create "a morbid state of the brain fibres concerned."[51] But he expects that in twenty years, having established "the great fact of the physical basis of the intellect," "the mental physician will be able to extract a specific recollection from the memory as readily as the dentist pulls a tooth, and as finally."[52]

Heidenhoff acknowledges some further problems with his process, admitting, for example, that there may be "shreds and fragments of ideas, as well as facts in his external relations" which are unaccountable without the eliminated memory.[53] But these things are no more a problem than the challenge faced by a drunk or a sleepwalker who has done things he no longer remembers.[54] Heidenhoff also acknowledges that the patient will know that he forgot *something*, which may create "slight confusion."[55] We see both problems arise in Madeline's case, and observe her filling in the blanks by creating plausible stories to account for matters which otherwise do not make sense to her.

Dr. Heidenhoff believes that his process has wide-ranging moral implications. On Henry's first visit the doctor notes that he finds patients generally come to him out of

> very genuine and profound regret and sorrow for the act they wish to forget. They have already repented it, and according to every theory of moral accountability, I believe it is held that repentance balances the moral accounts. My process, you see then, only completes physically what is already done morally.[56]

In this respect, he claims superiority to the "ministers and moralists" who preach forgiveness, but in fact leave the penitent subject to the ongoing tyranny of "remorse and shame."[57] Indeed, Heidenhoff notes, it is the most sensitive moral natures, those precisely who most deserve that their repentance relieve them of their misery, who are most likely to be tormented by the memory of their wrongdoing:

> The deeper the repentance . . . the more poignant the pang of regret and the sense of irreparable loss. There is no sense, no end, no use, in this law which increases the severity of the punishment as the victim grows in innocency.[58]

In contrast, "I free him from his sin" entirely.[59]

Heidenhoff is not content to beat religion at its own game; he believes that understanding the physical basis for memory and intellect, hence of all human actions, leads to the need for moral revaluation even beyond the specific consequences of his own technique. Some people, he says, may mistakenly think that sin must have painful consequences, whether physical or mental. They may believe that Nature is supposed to punish physical "vice and violence" with "diseases and accidents," and God is supposed to provide "moral retribution" via shame and sorrow. But in fact, both forms of retribution are essentially "blind, deaf and meaningless."[60] Nature does not care and Providence is unreliable, so human beings must step in. Just as no doctor would refuse to set the broken leg of a drunk, just to make an example of him, so we should be equally willing to relieve mental suffering by forgetfulness.[61]

Heidenhoff holds that because all suffering ultimately has the same physical basis, all suffering can and ought to be relieved. But in a manner familiar to arguments today about the difficulty of distinguishing between curing disease and enhancement, Heidenhoff's assumptions lead him to speculate that the world might be better off "if there were no memory" at all.[62] For without memory, we would still have "congenitally good and bad dispositions," but the bad would not, as they do, grow "depraved" by the demoralizing effects of memory.[63] "Memory is the principle of moral degeneration. Remembered sin is the most utterly diabolical influence in the universe. . . . more sin is the only anodyne for sin, and . . . the only way to cure the ache of conscience is to harden it."[64]

Heidenhoff's apparent insistence on the objective existence of sin will doubtless strike oddest those who most share his materialist assumptions, but in response to an objection by Henry, he develops his argument in a fashion that suggests it is not sin which is the problem, but the *idea* of sin. For Henry argues that even without memory, people will still be just as responsible for their acts as with it. Heidenhoff agrees, but with a twist: "Precisely; that is, not at all."[65] For "human beings are not stationary existences, but changing, growing, incessantly progressive organisms, which in no two moments are the same."[66] Hence, while crimes can be punished for prudential reasons—as a matter of "public policy and expediency"—they can never be punished *justly*, for it is never just to punish people for what they cannot help.[67] Why should I be punished for what I did yesterday, when I am not the same person who I was yesterday, and the person I am today has no control over what the person I was yesterday did?

In short, for Heidenhoff the self only exists moment to moment; by disconnecting the present from past and future the self becomes extremely thin. The moral implications of this thinness are clarified when Henry points out what he believes to be a flaw in Heidenhoff's argument. A person holding Dr. Heidenhoff's views would never need to erase any memories, since such a person would never feel guilty or remorseful about an act committed by an earlier version of himself. The doctor agrees that his process is indeed only for those who do not have the strength of mind independently to "attain" his philosophy.[68] In fact, we are born anew each day, each moment. "Is there not sorrow and wrong enough in the present world without having moralists teach us that it is our duty to perpetuate all our past sins and shames in the multiplying mirror of memory," the doctor exclaims.[69] Just as it is "only fools who flatter themselves on their past virtues, so it is only a sadder sort of fools who plague themselves for their past faults."[70]

In sum, Dr. Heidenhoff's process reveals that when we understand the true physical basis for memory, hence for human thought and action, we see how we make two errors when we fail to forgive ourselves or others: the failure to forget an irrelevant past and the failure to realize there is no continuous self to grant or receive forgiveness. That our conventional moral precepts are not based on these truths has the perverse consequence of making us either more vicious or less happy.

MEMORY AND MORALITY

Does Bellamy intend that we should test the adequacy of Heidenhoff's morality? For surely from the start we can wonder whether, as Heidenhoff believes, his process really exposes serious confusion in our existing moral intuitions on their own terms. Is there really a weak moral case for punishing a person who has forgotten his crime? For example, when someone in a drunken stupor does not recall the hit-and-run accident, we do not only punish him in order to deter others but because there has been a wrong done to the victim that exists whether or not the perpetrator recalls it. Or again, if you were told that someone who gravely wronged you could not be punished because he had his memory erased, it is not obviously unreasonable to be concerned both about the injustice of getting away with a wrong and the further injustice that the wrongdoer is allowed to forget it. Heidenhoff has not shown that such responses are confused in themselves; they are only confused if we are prepared to join him in rejecting the continuity of our personality, and hence our moral responsibility for past deeds.

But on this very point there would seem to be confusion on Dr. Heidenhoff's side in the matter of changing personality versus settled dispositions. Madeline clearly has a flirtatious and self-willed disposition. If all memory of her errors created by that disposition has been sponged away so that she cannot become, as the saying goes, "sadder but wiser" through them, surely she will constantly repeat them; being born anew will not free her from anything but rather would be the constant return of the same. On the other hand, Harrison Cordis, the man who wronged her, could be said to have attained the heights of Heidenhoff's "that-was-yesterday" philosophy, and is arguably the freer to indulge his dispositional lusts—his freedom to act in accordance with his dispositions is hardly a moral improvement. The moral wrong of Harrison Cordis's seduction of women would surely not be reduced even if the memory of the seduction can be wiped away.

Finally, take the case of Henry, who acts rather nobly, even if self-interestedly, in seeking the fallen Madeline out and offering her the shelter of wedlock. If Heidenhoff is correct and the memory of wrongdoing contributes to the worsening of bad dispositions, does memory of good deeds like Henry's also contribute to the improvement of good dispositions? Heidenhoff seems to look forward to a world where there would be no need to comfort the afflicted, hence a world with reduced opportunity for the development of other-regarding virtues.

BRAIN-DAMAGED GOODS

Beyond the problems we have just noted with Heidenhoff's argument, the 2004 film *Eternal Sunshine of the Spotless Mind* is nearly a point-for-point reexamination of some of his key premises, suggesting further grounds for skepticism about their adequacy. The story is not complex, but the movie presents it in a particularly thought-provoking way, such that the viewer only gradually comes to understand the sequence of events. The summary that follows, therefore, is very much a spoiler.

Joel and Clementine (played by Jim Carrey and Kate Winslet) meet, fall in love, and then fall out of love. To forget their painful breakup, Clementine goes to Dr. Mierzwiak (Tom Wilkinson) to have her memory of Joel erased; when Joel finds out what she has done he attempts to do the same. An underling of Mierzwiak's named Patrick (Elijah Wood) falls for Clementine when he participates in her memory erasure. Mierzwiak's process, as we will see, gives him access to Joel's diaries and other effects, and Patrick uses this information about

Clementine to court her. She finds his efforts confusing and increasingly disturbing for reasons we will speculate about later. Mierzwiak's secretary Mary (Kirsten Dunst) and his main technician Stan (Mark Ruffalo) are likewise involved, despite the fact that Mary carries a torch for Mierzwiak.

The situation comes to a head during the procedure to delete Joel's memory. Patrick goes off to be with Clementine, allowing Stan and Mary to get stoned and have sex while waiting for Dr. Mierzwiak's process to do its work on Joel. Midway through the procedure, Joel changes his mind about erasing his memories of Clementine, and we see inside his head his heroic mental efforts to try to preserve them. Stan notices that things are going wrong, and calls in Mierzwiak to solve the problems created by Joel's resistance. At the end of a hard night of erasing Joel's memories, Dr. Mierzwiak and Mary kiss, observed by the doctor's suspicious wife (Deirdre O'Connell)—who reveals when she confronts him outside Joel's apartment that Mary and Mierzwiak have already had an affair. Mierzwiak is forced to admit to Mary that she had agreed to have her memories of their affair erased. Angered, Mary mails Mierzwiak's patients their files. Joel and Clementine, who by chance or fate have met again and are falling for each other again, listen to the tapes detailing their previous unhappiness with each other. The movie ends as they are deciding to give their relationship another try.

We begin by observing that Mierzwiak tries to overcome what Heidenhoff acknowledged as imperfections in his process. To reduce confusion created by waking up in a doctor's office, the procedure is done in the patient's home at night. There will be no memory of having a memory erased. Mierzwiak also has the patient purge his life of any material reminders of whoever or whatever is to be erased, which is how Patrick gets his hands on Joel's Clementine memorabilia. Finally, to reduce further the chance for confusion based on forgotten events, Mierzwiak sends out postcards to inform the patient's associates not to talk about what the patient had erased.

Yet the movie demonstrates several plausible ways in which the process remains imperfect (after all, "technically, the procedure itself is brain damage"[71]). For example, we can understand why Joel would turn in a coffee mug with Clementine's picture on it, but how does he get away with not getting rid of the foldout bed that figures prominently in so many of his memories of her? Another problem created by the association among memories works the other way. Joel loses all memory of the cartoon character Huckleberry Hound apparently due to the association with the song "Darling Clementine." Or again, he is evidently a bit confused by waking up in the new pajamas that he bought for the procedure, but does not give them much thought. More troubling is the car damage he discovers the morning after the procedure. Clementine crashed it before their last fight, but, failing to recall that, Joel must construct a plausible story for himself that accounts for the now-unexpected dents. So he is upset but not confused.

All these examples fade in significance compared with Joel's discovery, inveterate diarist that he is, that two years of diary entries are missing. Surely it is only because he is so emotionally withdrawn in the first place that he is not more consciously concerned to understand this inexplicable loss. By contrast, the acute distress which Clementine increasingly feels, while doubtless exacerbated by Patrick's mimicking of Joel, shows another possible result of losing memories, more or less incompletely, without conscious awareness that one has lost anything at all. In short, there does not seem to be a consistent ability to predict what memories and associations will create problems when lost, or what exactly will be lost—hardly surprising given the rich associational and emotional character of our memories. Joel and Clementine have shared the *world*, and they cannot turn the whole thing over to Mierzwiak. Perhaps it is the fact of such ongoing associations along with the habits created by their dispositions that draws the two of them back together even after their memories of each other have been erased.

DOCTOR KNOWS BEST

The efforts taken to make the patient forget that he has forgotten something may therefore be of only limited utility from the point of view of the patient, but they *certainly* are useful to Dr. Mierzwiak. We see one hint of how when we find out that the "don't ask about . . . " postcards about Clementine's procedure are sent to everybody but Joel. Had Joel (as intended) remained in the dark about Clementine's decision he would not have stormed into Mierzwiak's office as he did when he found out. In like manner, if Clementine grows increasingly unhappy and confused as Patrick courts her, she will not see herself as an unhappy *return* customer in Mierzwiak's office, although she may see him again as a "new" customer, unhappy for reasons she cannot connect to him. In short, Mierzwiak's improvements, while nominally for the good of the patient, in fact insulate *him*, to the extent they work, from any bad consequences of his procedure. The postcards even serve the additional useful function of providing advertising when, of necessity, there can't be direct word-of-mouth.

Mierzwiak has arranged his process so that as much as possible he will not be confronted by the problematic consequences of his work. The irresponsibility inherent in his methods is reflected yet more clearly in the behavior of his underlings. Drinking, drugs, theft, and perhaps sex seem to be normal activities as they "monitor" the memory-erasing procedure. The forgetfulness they induce in others allows them to forget themselves, professionally speaking; they can satisfy their transitory desires with impunity. Now of course, in any care-giving situation there is a possibility for abuse such as this, and presumably it only increases when the patient is unconscious. But Mierzwiak's employees are further liberated by the fact that those from whom they steal, for example, will not even remember they were unconscious and under treatment. Joel finds he has less whiskey than he expected in the final moments of the film—but even knowing what he knows by then he does not connect it with the procedure.

The technical abilities and limits of Dr. Mierzwiak's process provide fresh opportunities for human failings, which brings us closer to the problems the film suggests with the moral case that Heidenhoff makes for memory erasure. It is easy to connect the irresponsible behaviors we have been looking at with Heidenhoff's denial of any continuity of self that would justify moral responsibility. Just about everybody in the movie has the very thin sense of self that Heidenhoff believes is a great moral achievement. Stan's stock humor is the quick personality shift from brusque professionalism to clownishness. The colorless Patrick readily adopts Joel's persona when given a chance, turning it on and off as needed. Mary gets her opinions from the multi-voiced *Bartlett's Familiar Quotations*. Clementine laments that she gets her personality out of the pastes for her ever-changing hair color. These thin identities are what it means to "attain" to Heidenhoff's philosophy of *that was then, this is now* and *the I of today is not the same as the I of yesterday*.

If the results are bad all around, Dr. Mierzwiak himself illustrates the dangers best. In the movie he is not played as a mad scientist or evil genius; indeed his fatigued diffidence and calm contrasts favorably with Heidenhoff's preening iconoclasm. Joel barges into Mierzwiak's office, he calmly sees Joel. Mary kisses him, he kisses back. Because his wife Hollis was upset, he reasonably arranges to have Mary's memory erased. (Evidently he could not get Hollis to agree to the same thing.) But Hollis thinks he is a monster, and there is something to her point of view.

We see how when we think more about how he erased Mary's memories of their previous affair (in the shooting script, there is also an abortion involved).[72] While subjecting her to the treatment was a decision he claims they came to together, we have no reason to think, given how the procedure works, that this claim would have to have been true. And even if it were, the obvious asymmetry in their relationship makes the claim of a mutual decision suspect. He could have employed against her the very fatherly/professional authority which is clearly part of what attracts her to him.

But Mary and Mierzwiak's case exposes a yet deeper problem in

Heidenhoff's belief that we lack moral responsibility for our pasts while maintaining that we have dispositions that determine the general direction of our behaviors. In what proves to be a particularly poignant speech, Mary articulates the Heidenhoff-like hopes behind Mierzwiak's work. "To let people begin again. It's beautiful. You look at a baby and it's so pure and so free, so clean. Adults . . . they're like this mess of anger and phobias and sadness . . . hopelessness. And Howard just makes it go away."[73] It is true that Howard can make Mary's memories of their affair go away, memories presumably painful mostly in retrospect. But *everything* does *not* go away. Instead, he "frees" her to fall in love with him again and creates a fresh opportunity for more infidelity. Mary cannot learn from her mistakes; her unhappiness over his unwillingness to leave his family for her will not cause her to grow angry with him, or disillusioned, or to leave her job and move on, because she has forgotten she was unhappy. On the other hand, how satisfying it must be for Mierzwiak to have her prove his power over her by falling in love with him yet again.

The same dynamic is at work between Joel and Clementine. From a purely prosaic point of view, if all had gone as expected their mutual erasure would simply have opened the door, should they meet again, for them to repeat all of their past mistakes. The same things would attract each about the other, the same things would eventually drive them crazy. Mary's intervention in sending out the patient files[74] opens the door to something new and different in their relationship, but this *deus ex machina* paradoxically merely restores the possibilities inherent in normal life: "boy meets girl, boy loses girl, boy gets girl." At the movie's end, for Clementine and Joel to reunite they must forgive each other for doing and saying things they have no memory of doing and saying. Granted, that may be an easy kind of forgiveness. But some forgiveness would have been possible had they simply resolved to get together again after a normal breakup—which is not to say they have much of a future in any case.[75]

Against this argument that *Eternal Sunshine* represents the problems created by Heidenhoff's notion of a thin moral self, it could be argued that it is in the process of getting his memory erased that Joel comes to *know* himself. He attains the Proustian insight that if he truly no longer loved Clementine he would not feel the need to show her he no longer loved her,[76] which is, in his own mind, what he wanted to demonstrate by the erasure. Of course, the effort really makes no sense at all since he cannot demonstrate anything to someone who no longer remembers him; he can only demonstrate it to the Clementine of his memories, who will likewise be gone soon enough. Furthermore, such an insight is hardly useful if he cannot remember it; Joel's already thin self becomes all the thinner by his decision to deal with the problem of Clementine by erasing her memory.

Still, Joel's motive gives the lie to Heidenhoff's claim that patients will come to him out of sincere repentance. We see Joel make his decision on the basis of that combination of anger and sorrow and desire for revenge which is not atypically a result of failed relationships. (Clementine, typically for her, is said to have undertaken the procedure impulsively.) Where Heidenhoff thought no one could want to revenge himself on an ignorant wrongdoer, Joel exhibits animosity not in spite of but *because* he knows that Clementine no longer has any knowledge of the wrongs he thinks she has done him. Joel's revenge is typically passive-aggressive; we are reminded of how fortunate that is by the disturbing fact that in happier days he and Clementine apparently played murder/suicide games together, their thin souls toying with complete dissolution. Were Joel's "disposition" different, Clementine's choice to undergo Mierzwiak's process could have resulted in tragedy rather than romantic comedy.

Finally, Heidenhoff's own argument about the moral meaninglessness of psychic pain is shown by the movie to lower the bar for what kind of pain is tolerable. Perhaps it is simply moral progress that sexual relations outside of marriage no longer ruin a woman's life. But at least Madeline is portrayed as seeking out drastic measures because her life was indeed, by the standards of the day, ruined. Is it equally progress when one employs such drastic measures for a rather routine breakup such as Joel's and Clementine's—or, as we see with another patient, for the loss of a dog? Or is it rather once again an indicator of the difficulty these thin selves have in finding any meaning in their own actions and choices?

Nothing could stand in sharper contrast with the attitudes we see playing out in the movie than the poem from which the film draws its title, Alexander Pope's retelling of the great medieval love story of Peter Abelard and Eloisa in "Eloisa to Abelard" (1717). The poet has Eloisa both lament and celebrate the impossibility of her ever forgetting or overcoming her love for Abelard; "Tis sure the hardest science to forget." The nuns around her may enjoy the "Eternal sunshine of the spotless mind," but with her own "Far other dreams," such enjoyment is as impossible for her as it is undesirable.[77] Unlike nearly everyone in *Eternal Sunshine*, she knows who she is, knows whom she loves, and is willing to live out the consequences to the last.

SHRUNKEN SELVES

What can we learn from these fictional sources about the moral challenges we may face in a world of transhumanist enhancements? Enthusiastic advocates of enhancements don't like to think seriously about the practical inevitability that the enhancements will work imperfectly, and that those technical imperfections will combine with human imperfections (for we will not make decisions to be enhanced from an enhanced position) to create opportunities for people to take advantage of one another. That this is not a new problem does not mean it is not a serious problem. To the contrary, its persistence over time exposes the utopianism of any who think that the *sine qua non* for the proper deployment of enhancement technologies is more technology plus the maximization of individual freedom of choice.

These fictional works remind us that enhancements are not, as techno-libertarianism would have us believe, just a matter of what one individual chooses to do or not do. The choices that become available will arise out of the old world, with all of its unenhanced (or, at later stages, less enhanced) baggage and, as we saw in *The Diamond*

Age, the choices will be conditioned by those circumstances. The assumptions behind thinking of some change as an enhancement—why we think it is desirable in the first place, what we think it will tell us about ourselves—will have an effect not just on the enhanced, but on their relationships as well, both with their enhancers and with all others who form a part of their world. From this point of view, to imagine as do the transhumanists nothing beyond the greatest possible freedom to make choices among the widest variety is really a restatement of the morally thin conception of the self. That observation in turn casts light on the transhumanist ideal of endless self-transformation. Enhancement is no longer a matter of becoming the best one can be when there is no core or stable self to enhance; perpetual change is equivalent to destruction.

Dr. Mierzwiak and his people act out the consequences of Dr. Heidenhoff's denial of any continuity of self that would produce moral responsibility in what is by transhumanist standards a quite prosaic world of still limited choice. Imagine how much worse their behavior could be when they have, perhaps like Daphne, less reason to think of themselves in terms of a self that could bear moral responsibility. After all, the whole notion of enhancement is predicated on a restless dissatisfaction with what we are, in the name of the as-yet-unknown possibilities for what we might be, and the hope for discontinuity between what evolution has produced by chance and what intelligence can manage deliberately. But from there it is but a short step, already taken by some transhumanists like Ramez Naam, to casting doubts on the existence of a self altogether—the self is, some would say, mere "user illusion."[78]

There is some irony, then, in defenses of enhancement in the name of liberty. To be a truly free human being, to make one's own choices, implies accepting responsibility. It makes sense to advocate liberty if there is such a strong, responsible, moral self waiting to be free. But if, following Heidenhoff and today's transhumanists, we think that there is no such morally responsible self, we are left with only the willful or passionate choices of the moment, rationalized or otherwise, the expression of which is hardly liberty at all but license. There are of course huge moral hazards in a nanny state that would presume to protect the Joels and Clementines of the world from the selfish ministrations of the likes of Mierzwiak and his minions. However, in the transhumanist world of the right to choose without responsibility, the power that our future Heidenhoffs and Mierzwiaks will have over nature, and the power that they will therefore have over human beings, will be not so much the trump card as the only card to play when doubts arise about where our knowledge is taking us. Paradoxically, the world recreated by human beings will be predicated on the law of the jungle; our technological might will make right.

It may well be that this situation arises out of the fact that transhumanism is an effort to maintain some concept of progress that appears normatively meaningful in response to Malthusian and Darwinian premises that challenge the idea of progress. Malthusianism has come to be defined by thinking that the things that appear to be progress—growing populations and economies—put us on a self-destructive course, as we accelerate toward inevitable limits. But it almost seems as if, in the spirit of Malthus's original argument, there is something inevitable also about that acceleration, that we are driven by some force of nature beyond our control to grow until we reach beyond the capacities of the resources that support that growth. Meanwhile, mainstream Darwinian thinking has done everything it can to remove any taint of progress from the concept of evolution; evolution is simply change, and randomly instigated change at that.

Transhumanism rebels against the randomness of evolution and the mindlessness of a natural tendency to overshoot resources and collapse. It rejects, as we have seen, the "assumption of mediocrity" in favor of arguing that man has a special place in the scheme of things. But its rebellion is not half as radical as it assumes, for transhumanism builds on the very same underlying conception of nature that the Malthusians and Darwinians build on, vociferously rejecting the thought that nature has any inherent normative goals or purposes. While it rejects blind evolution as a future fate for man, it accepts it as the origins of man. While it rejects a Malthusian future, it does so with threatening the same old apocalypse if we do not transcend ourselves, and, in the form of Kurzweil's law of accelerating returns, it adopts a Malthusian sense that mankind is in the grip of forces beyond its control.

Because transhumanism accepts this account of nature, it is driven to reject nature. Rejecting also any religious foundations for values, then, it is left with nothing but socially constructed norms developed in response to human power over nature, which, given the unpredictable transformative expectations they have for that power as it becomes not-human, ultimately amounts to no norms at all. Transhumanism is a nihilistic response to the nihilism of the Malthusians and Darwinians.

This moral lacuna at the heart of transhumanism is why it can advocate the progress of dehumanization so enthusiastically. Bostrom may assert that posthumans can be humane, but human morality is built on human capacities and the circumstances of human life. If these are precisely the things that are going to change radically, it is hard to see how the progress of transhumanism can really be called progress anymore. At least, if the Singularity is the outcome, transhumanism cannot be "progress" understood as achieving or approaching some humanly comprehensible ideal or goal. It is progress only if a leap into the unknown can be called progress. That being the case, the only thing that can be said for it by anyone other than someone who enjoys taking incalculable risks is that it is *necessary* that we go down this path.

But since we cannot in fact know if it is necessary, the Singularity represents an aspiration, and aspiring to it has consequences long before it arrives, if it will arrive at all. To aim at the Singularity is to aim at something without content, or (which amounts to the same thing) to aim at any content at all. Such a goal is peculiarly fitting to the thin, willful self that we have seen come to the fore within the transhumanist framework, a self that supposes itself capable of any content but can only prove that capability by serial negation. This self is the apotheosis of what the philosopher Thomas Hobbes had in mind as a defining human characteristic, the "perpetual and restless desire of power after power, that ceaseth only in death,"[79] except of course it now expects to cheat death as well. We saw how Moravec recognizes that there is something deceptive about the transhumanist promise of immortality, because the "I" that seeks it will not survive its own desire for constant transformation. In fact, the same deception is at work in all of transhumanism's promises to be looking toward solving human problems or other benevolent representations. However sincere these promises, the good intentions reference a world whose destruction in its present form transhumanists are at the very least reconciled to, if not actively seeking. As they see it, the "present form" is mere accident, unworthy of the respect of intelligent beings.

There is of course something to this line of argument. Our desires are restless, and hard to control. Genuine Stoic resignation about the ills of the world has only ever been genuinely possible for a very few. It becomes all the more difficult in a world where it is manifestly possible to meliorate ills that once called for resignation, even if so doing has been known to create new problems. In any case, change seems to be in our blood. Those who even try to use the rhetoric of technological stasis, let alone rollback, are few and far between, and it is yet harder to find those living in a manner consistent with such beliefs.

So surely the safest bet is that the world will continue to change in response to human activities. Even so, the transhumanists may or may not be correct that all roads that are not immediately disastrous lead to the Singularity. This uncertainty means that a great deal is at stake respecting the extent to which their vision of the future is adopted. If it becomes the dominant narrative of technological change and scientific progress (all technical questions aside) it becomes more likely by that very fact. That would be unfortunate, given what we have seen in this chapter. The transhumanists have not adequately dealt with the very thin moral self and the problematic relations between enhancers and enhanced that are central to *Eternal Sunshine*. The disparity in power that would be problematic under any circumstances is all the more so in the time of the thinning self. The will can become all the more willful when it is no longer even directed toward what I want to achieve, which after all implies some kind of self and limit, but is instead liberated to be all-negating. John Harris, avoiding as is his wont any use of transhumanist terminology as he makes his transhumanist arguments, tells us proudly that he does not "recognize finitude, only the limitless possibilities of the human spirit and of human ingenuity."[80]

Finitude is, of course, all around us and in comparison to this reality transhumanism offers what is ultimately a kind of nothingness, be it in the form of mere negation or the unknowability of the Singularity. From this point of view, limits don't look so bad. If we go back to the common-sense beginnings of transhumanism in the admittedly extraordinary potential that modern science and technology hold, how might we begin to think about a future within limits—a *human* future?

CHAPTER FIVE

The Real Meaning of Progress

PROLOGUE: FATHERS AND SONS

I've never really given that much thought to birds. Eating them, of course. But beyond that they do what they do and we do what we do. Dad's another story. If it hadn't been "let's try this experiment to escape from Minos," I'm guessing that sooner or later he would have found some other excuse to strap on wings. That's just the way he is—which means never satisfied with the way he is.

It's an admirable enough trait. I guess. We're supposed to strive for excellence, at least as much as the gods allow. But excellence for Dad seems to be more doing something new than doing something well. I've seen the "wings" he's making, and I'll never strap them on sober, I can tell you that. It's not that I'm against new things—where would we be without Prometheus? But Dad thinks that the fact that he *can* do something is reason enough to do it. So let anybody offer him the chance to do something new and he is all over it. No thought for the consequences. Even now I'm pretty sure that he misses the irony of the fact that we would not need these stupid wings if he had not made that disgusting hollow cow. And this damn maze.

If we do get home after all this I'm going to have to talk with him. It's not that I think he is likely to change, but the gods know that somebody has to try and straighten him out before something really terrible happens. I'd hate to be the one to have to pick up the pieces.

When appealing to common sense, transhumanists promise a better world in humanly comprehensible terms. However, their own assumptions lead them to abandon those promises in favor of willful change toward incomprehensible outcomes. They promise that science and technology will provide us with more of whatever it is we want more of at any given moment—and it is indeed hard to deny the attraction of that promise to people like us. The hitch is that people like us are not going to be around to enjoy it. Indeed, we are not even supposed to see our elimination as a cost at all, but as a great benefit. At least, for the transhumanists, this outcome is in some fashion necessary, and we are supposed to accept that technological might dictates right.

The transhumanists believe that their ideas represent *progress*—not just technological progress, but progress in the much larger sense of humankind fulfilling its ultimate destiny (a destiny of overcoming itself). Now, one might have thought that ethical guidance would be central to deciding whether a given discovery or invention actually served to advance the well-being of humanity in a way that would deserve labeling it progress rather than mere change, or worse. When Condorcet spoke of progress, for example, we could see his hopes for moral improvement. But shorn of any serious moral content, the measure of progress—if it can be said to exist at all—comes to be simply our amazement at, or dissatisfaction with, the present state of our discoveries and inventions, our awed anticipation of what might yet be achieved. Indeed, our terror about what might go wrong along the way becomes a kind of measure of progress.

The result of framing the question of progress in this way is obvious in popular discussions of the future of science and technology. First, start with a little history to produce an attitude of pride that we know so much more than we once did. Then look at what we know now, and stress the dangers of our remaining ignorance. Finally, anticipate future discoveries, combining hopes and fears for what might happen with a humbling sense that, if only we stick with it, those who follow us will know more than we do and be able to do far more than we can. It would look like Winwood Reade's vision of mankind "ripening towards perfection," if it had the idea of perfection and Reade's tragic insight into the sacrifices involved.[1]

Beyond that, "progress" becomes the sheer accumulation of information, a kind of hoarding mentality that is based on the belief that you never know what might come in handy someday. This helps to explain the widespread belief that any effort to restrain science or technology on the basis of ethics represents a threat to progress. After all, if progress is mere accumulation, then of course restraint *is* a threat. Yet to see this concern as simply expressing the self-interest of researchers and inventors is to do them an injustice. Haldane wrote that the prospect for the future "is only hopeful if mankind can adjust its morality to its powers,"[2] and many well-intentioned professionals probably would agree that it is the job of morality to adjust to scientific and technological change without appreciating how that amounts to saying that might makes right, since they are saying that the question of what we ought to do must always bow to what we have the power to do. Others probably believe that ethical restraint is a weak reed. That is why we frequently hear the argument that if "we" restrain ourselves with respect to some line of research or development, "they"—some other country, usually—will not, putting us at a disadvantage. Combine these two perspectives and you can begin to understand why so often in practice "ethics" of science and technology becomes a matter of filing the right paperwork, following professional codes, publishing in highly specialized journals, or scientists and engineers being willing to have meetings with people from otherwise safely segregated humanities departments to talk about the "ethics of . . . " this or that technology or line of research.[3]

In fact, however, the kneejerk suspicion of any effort to limit developments in science and technology represents something of a betrayal of the bargain that gives science its high place in the modern world. The enlightened acceptance and public encouragement of science and technology was built on the assumption that freedom for such research and development by those so inclined would serve *human* well-being. The so-far largely successful results of that bargain have made the modern world what it is. Whether the bargain can be as successful if we lose sight of its terms is another matter, about which one is entitled to be quite skeptical.

Some may argue, in the manner of this book's epigraph from Albert Einstein, that today we are too sophisticated to think that there really is any rational answer at all to the question of what human well-being means; hence it could hardly help us think about how to limit our growing power over nature. Cultural and moral relativism, historicism, postmodernism, dogmatic materialism, and fashionable nihilism all create obstacles to taking the question of the human good seriously in our time, obstacles that were not created by advocates of the eclipse of man but that are consistent with their unwillingness to tackle moral questions. But if it is no betrayal of the bargain behind our scientific and technological prowess to at least *inquire* into the limits represented by the pursuit of human well-being, is it not all the more urgent to pose the question when the ultimate promises of transhumanism so blatantly reject that goal in the name of powerful enhancers, willful negation, and the mystification of the Singularity?

It is no new observation that the great increase in our powers coexists with a diminished capacity to think about them with any kind of moral realism.[4] But slighting ethics does not genuinely serve the cause of science and technology, since they only matter in human terms if they truly serve our humanity. When progress is defined by dehumanization, it is obvious that this result is by no means guaranteed.

While transhumanism is still a fairly recent development, questions about the extent to which human ingenuity serves human beings well are hardly new—as witness the ancient Greek myth of Daedalus and Icarus. The details of the story are familiar: Daedalus is a great craftsman and engineer, the builder of the Labyrinth used to entrap the Minotaur on the island of Crete. Despite this service, Daedalus and his son Icarus are imprisoned on Crete by King Minos. Daedalus fashions birdlike wings of feathers and wax so he and his son can fly to their escape. Ignoring his father's cautions, Icarus flies too close to the sun; his waxen wings melt and he plunges to his death in the sea.

The tale has drawn a fair amount of attention from artists over the centuries, including the three paintings we will discuss in the next section. The discussion that follows does not aim at offering a comprehensive account of the human good. Nor does it attempt to defend any particular limit on how we might use science and technology to preserve a future of *human* well-being. But it suggests how we might begin to think about such limits in the course of even modest reflections on the world we are making day to day with science and technology. The transhumanist arguments obscure what is present in front of us in this world; its imperfections and failures, for example, are swept away in a tide of technological determinism drawing us on to some distant horizon of imagined possibilities. That transhumanist farsightedness is then taken to be the best framework by which to give a trivializing and dismissive meaning to present-day things. The three paintings we now turn to provide illustrations of the range of moral responses to the eclipse of man, responses that can illuminate the reasons for rejecting transhumanist farsightedness and put us in a better position to take seriously the human purposes that science and technology promised to serve.[5]

WINGS OVER THE WORLD: THREE VISIONS

Our first image is "Daedalus and Icarus" by the French writer and painter Charles Paul Landon, who would eventually become better known as a writer on art than an artist. It is a marvelous illustration of the transhumanist hopes for the progress of inhumanity. The wings seem so natural—the thin fabric straps that bind them to the bodies of Daedalus and Icarus are easy to miss—that the pair look less like human beings wearing improvised wings than like winged beings. We see a moment of great promise, aptly in what seems the light of a new day, as Icarus steps into his first moment of flight, with the intent assistance—perhaps a slight push, perhaps just steadying hands—of

Charles Paul Landon, *Daedalus and Icarus*, 1799
Musée des Beaux-arts et de la Dentelle, Alençon
Picture: David Commenchal

his ingenious father. How Icarus feels is anyone's guess (does his face need to be quite so heavily shaded?), and Daedalus exhibits more concentration than amazement or even satisfaction at what he has achieved. Without clothing, with only a sky as background, and with only the vaguest of classical motifs in the pedestal on which they stand and perhaps in their hair, the picture presents this moment of accomplishment, the dawn of a new day, almost completely abstracted from time, place, personality, and circumstance (Icarus is even curiously androgynous). Thus, all the distinctions by which we normally define human beings, except the one that highlights our ingenuity, are missing, and perhaps therefore arbitrary, and the painting becomes a pure tribute to the magical-seeming potential of human invention. Landon of course knows we know what happens next in the story. But from this starting point, it is almost impossible to believe that Icarus will fall—it seems at least as incomprehensible as the fall of some angel. It is as if Landon has in mind a new version of the story, in which father and son both survive.

Our second image, "The Death of Icarus" by the contemporary German painter Bernhard Heisig, is a powerful illustration of a certain kind of problematic critical response to the eclipse of man. In contrast to Landon's painting, Heisig's shows the end of the story, perhaps at sunset, with a screaming, terrified Icarus (apparently a self-portrait of Heisig[6]) the center of attention as he is crucified on the obvious artifice of his father's inventiveness. The background echoes the famous painting of the Tower of Babel by Pieter Bruegel the Elder, while the prophetic, prominent pointing finger beneath Icarus' left wing could be a rotated image of the hand of God from Michelangelo's *The Creation of Adam*. All these Biblical visual references add up to a strong warning against overreaching technological ambition—a reminder of the human moral imperfection that conditions the way we use our amazing abilities, and a useful corrective to transhuman aspirations.

Yet Heisig, like Landon, simplifies the moral equation at work in the story. Despite or even because of its cultural quotations, the context in which he places the unfolding events is perhaps even more mysterious than Landon's. The violence of this image, something for which Heisig's work is generally known,[7] is quite shocking, and puts the primary focus on the very direct line between Daedalus' innovation and Icarus' terror. "Here is where your creative pretensions will end up," the painting seems to be saying: not just in failure but in horrifying disaster visited upon those closest to you. So in a curious way this deeply negative outlook depends on accepting the same kind of necessity that transhumanists like to claim drives their project. It

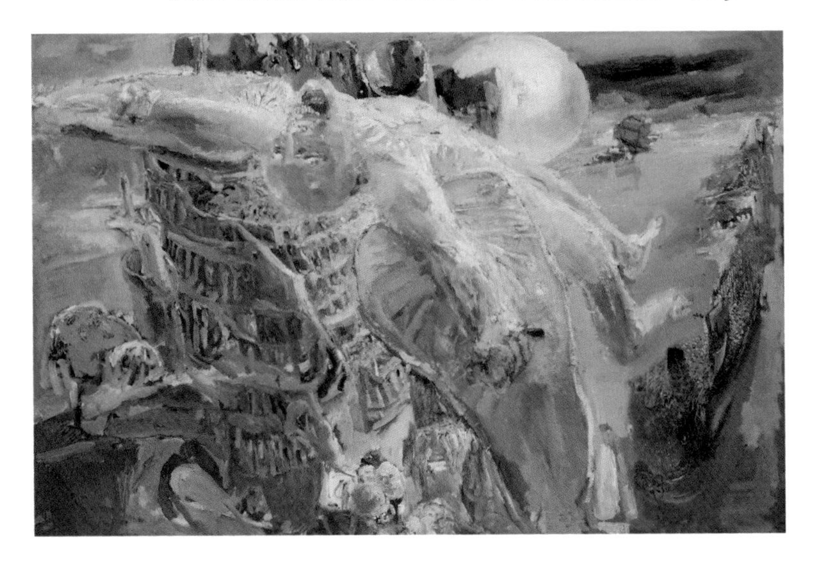

Bernhard Heisig, *Der Tod des Ikarus* ("The Death of Icarus"), 1979

appears that for Heisig failure is the only option given that human beings are as they are. There is no point in speaking of progress at all if, from Adam on, we are fated to make the same overreaching kind of mistakes again and again, as the painting seems to suggest.

However, that perspective overwhelms the fact that this son is, after all, not bearing his cross because he is self-sacrificing or even fated to be sacrificed, but is a young man personally responsible for having flown, against the warnings of his father, too close to the sun. Of course, we can hardly expect that sons will always obey their fathers. But it is not clear whether Heisig sees how the moral significance of Daedalus' work looks very different when we take account of the fact that his son's character plays a role in the way things turn out.

Our last picture is "Landscape with the Fall of Icarus," generally attributed to Pieter Bruegel the Elder. It is loosely adapted from Ovid's telling of the story in *Metamorphoses*. Much has been and deserves to be said about this remarkable painting; let us therefore just stick with

Pieter Bruegel the Elder, *Landscape with the Fall of Icarus*, CA.1555
Musees Royaux des Beaux-Arts de Belgique, Brussels, Belgium /
Giraudon / The Bridgeman Art Library.

the obvious. Icarus plays a very small visual part in the story the painting tells; it depicts nearly the same moment of failure that Heisig presents, but in Bruegel we see it at a great and impersonal distance. A splash at the bottom right, some scattered feathers, and Icarus' tale is done; Bruegel has to play with the perspective a bit even to make him as visible as he is.

Of course, that minimization is its own kind of warning against overreaching: who wants to be remembered most for self-imposed failure? But beyond disobeying his father's instruction to follow a middle course between sea and sun, what is the source of Icarus' failure? Ovid's version presents Daedalus in a not-very-flattering light that casts some doubt on his desire "to work on unknown arts, to alter nature,"[8] but it is not so clear that Bruegel agrees. The shepherd tending his flock in the middleground of the painting may come closest to living with nature unimproved (or at least, improved so long in the past as to *appear* unimproved), but the fisherman certainly requires artifice to make his living. And our farmer's contour plowing may reveal him to be conservation-minded, but his plow, clothing, and literal reshaping of the earth together suggest no small degree of technological sophistication, a suggestion only amplified by the great argosies plying the waves and the alabaster city in the distance that is their apparent destination. In short, Icarus is placed on a continuum of very human enhancements of the given. Yet the tiny part that Icarus plays in the picture suggests a very different view from that presented by Landon and Heisig.

Bruegel's potential witnesses to the fall of Icarus are barely witnesses, if at all—they are certainly not dismayed, as in Heisig.[9] In this respect Bruegel changes Ovid's story:

> A fisherman, who with his pliant rod
> was angling there below, caught sight of them;
> and then a shepherd leaning on his staff
> and, too, a peasant leaning on his plow
> saw them and were dismayed: they thought that these
> must surely be some gods, sky-voyaging.[10]

Bruegel keeps the same characters, but not one of them seems astonished by what he has seen; indeed, it is not really clear they have seen anything at all. Because we know what is happening, and because we know Ovid, we can wonder about their *lack* of wonder. Is it mere peasant stolidity, or the general human obliviousness in the face of something new and (as some would say) important? Perhaps the fisherman sees Icarus. He might have thought, "*Was that really a man falling from the sky? No, it couldn't be.*" before returning, like the farmer, to doing his job. A flying man is outside of his expectations, and a god would hardly crash-land. Our shepherd is a more interesting case; he, not the plowman, raises his eyes, but not in the right direction to observe Icarus' fall. Perhaps he looks up because he hears Daedalus calling to his son? Or perhaps gazing up at the birds he daydreams of flying? But for him it is a passing fancy; not everyone has the genius and motive of Daedalus to turn such dreams to reality.

In any case, the most thought-provoking thing about Bruegel's painting of the story is that he does not have us (as Landon and Heisig do) asking much about Icarus, does not have us wondering about his wondrous or terrible flight. By placing Icarus in a relatively commonplace landscape, Bruegel instead makes us wonder about what Icarus means to these commonplace figures. He is an easily missed part of Bruegel's story, overwhelmed by Bruegel's seductive landscape of the everyday. In other words, in the world out of which they arise, innovations by the likes of Daedalus may hardly appear at all. Daedalus' great purpose, to find a clever way of freeing himself and his son, and Icarus' great failure, echoing his own father's quest to be free, do not intersect directly with what is important to the others in the painting.

On one hand, Bruegel's painting reminds us that the mundane is not given once and for all: there was a time when ships, cities, plows, fishing rods, and even herding sheep were new, the kinds of novelties that might have been, like Icarus' strange flight, either dangerous or hardly worthy of notice by sensible people.[11] On the other hand, Bruegel's painting suggests the limits to the power of change, for the world he depicts would have been just as recognizable to a viewer four centuries before his time as it is for us four centuries after. Even four millennia before Bruegel there was food to grow, there were animals to catch or tend, there were goods to trade, and there were disobedient children. Familiarity across such time scales may be less than the blink of an eye from some imagined cosmic perspective, but this lived human experience provides the continuities that ground and shape human life. It is because of this backdrop that we have the chance to judge innovation and change to be merely that, or in some real sense *progress*; there has to be a human condition in order to speak of progressive improvements to it. Furthermore, it is because of this backdrop that the importance of an innovation will become obvious only in retrospect; we cannot know in advance whether its significance, if any at all, comes from the manner in which it is integrated into the old world, or overturns it. Bruegel stresses what Landon did not: that Daedalus' achievement was a dead end.

BETWEEN TWO VISIONS OF THE FUTURE

There is a further consequence of Bruegel contextualizing Daedalus' invention within the commonplace earthy, watery, and airy worlds of the farmer, fisherman, and shepherd. Can we not imagine them absorbed not merely in their work, but in those they are working to support? Their own aspirations for and worries about their families and communities would actually be their point of contact with the otherwise unfamiliar events whose final moments unfold before them. As they pursue their daily lives, Bruegel's figures look better dressed and better off than might have been true for their forebears; they might have similar hopes for their own descendants' material improvement without imagining them to be like gods. If they could meet Daedalus, then, we can well imagine they would be more inclined to marvel at his escape from a tyrant and commiserate with him about a disobedient son than focus on the details of his invention. Daedalus does not invent in a vacuum.

The existence of this shared world may be frustrating to those who today pride themselves on being at the scientific and technological cutting edge. But Bruegel reminds us that, practically by definition, the cutting edge is not where most people are; it does not even loom large in their lives. (Indeed, I know of no ancient story that suggests that Daedalus himself ever flew again; even for him flying itself was perhaps less an aspiration than escape.) Still, to those who think themselves in the vanguard, like today's transhumanists, the rest of us will appear as the fisherman, plowman, and shepherd—the ignorant and unobservant who through mere inertia ignore a fabulous future, or seek to keep it at bay. However, there is a certain falseness about this perspective, since in fact the vanguard does not really live in the future they imagine and must continue to rely on the existing world built by the mundane choices and motives that they would rather not acknowledge. Ovid presents Daedalus as an escaping prisoner, Icarus as proud and foolish. Such human details decisively shape how innovations come to be and how they are used. The extraordinary ambitions of the transhumanists and our other latter-day followers of Daedalus will be mediated by the complexities of the ordinary.

Is there any force that can move the ordinary world inexorably toward the radical changes foreseen by the transhumanists? Mightn't the powerful forces of Malthusian scarcity and Darwinian competition push us in that direction? Bruegel, as a man of the sixteenth century, of course cannot be blamed for being unaware of such ideas. Does his unawareness undermine the human vision that he is seeking to present?

The first thing that needs to be noticed in order to answer this question is that the lesson of ongoing scarcity Malthus taught, and to some extent also the lesson of Darwinian competition, is shocking only after hopes for *ending* it, such as those articulated by Condorcet, come on the scene and become widely adopted. Without a vision of progress like Condorcet's, scarcity and competition would more or less simply be taken as definitive of the way things are. You need to have a vision of progress first, and then the Malthusian challenge can contribute to its radicalization.

Bruegel chooses in this painting not to highlight the worst consequences of scarcity and competition, aspects of life that would have been perfectly obvious to a man of his time. But that does not mean they are absent entirely. We see them at work in the inequality implicit in the painting, the distinction between city and country life, the somewhat menacing island fortress, the occasion for Icarus to be flying that is the story behind the picture, the existence of trade, the fact that the three peasants must make their livings by the sweat of their brows. All such things are just the norm in Bruegel's world. His acceptance stands in stark contrast to the mental gymnastics that advocates of transhumanism have to attempt in order to deal with scarcity and competition. As we saw, Drexler is driven to reject Malthusian scarcity in the near term, accept it over the longer term, and then argue for its irrelevance even in the long term. To one-up him, Kurzweil has to speculate that things we take to be fundamental physical limits may prove subject to intelligent manipulation after all.

We are fortunate to live in a world where, based on some very hard work, Malthus's gloomy predictions about scarcity so far seem to have been confounded. Perhaps, then, Bruegel's resigned attitude is misplaced—perhaps innovation, like that of Daedalus, deserves a much more central placement in our picture of the world. But if scarcity and competition have not yet done their Malthusian worst, we should also note that we can hardly claim that Condorcet has obviously been vindicated. Material progress has had costs with respect to both want and excess that Condorcet did not anticipate, and to be satisfied that it has produced the moral progress he expected it to create would plainly be mere smugness. If, as suggested above, some sort of progress is not entirely alien to Bruegel's picture, it may navigate between the Scylla and Charybdis of these alternatives—between the pessimistic vision of Malthusian decline and the optimistic vision of progress—that so readily occupy our contemporary imaginations. Bruegel's in-between depiction may therefore be truer to the world we live in.

Finally, does Darwin's argument for the mutability of forms of life require us to abandon the perspective on human things that Bruegel seems to be suggesting? At the simplest level it does not, because a great many people in the world today, perhaps even most of them, are as little influenced by Darwin's ideas as Bruegel's peasants were, despite all the efforts of bestselling neo-Darwinian popularizers. We may decry such ignorance or (worse) obstinacy, but there it is; integrating Darwin into a horizon for understanding the world remains the achievement of a relatively small number, and it would be a brave soul today who would claim that these represent the vanguard the rest must necessarily follow.

Furthermore, it is not really that clear that Darwin's vision of natural evolution should be *expected* to change the shape of our lives to something other than what Bruegel observed. What life lessons ought we to learn from Darwin?[12] If his work represents a warning against a certain kind of human pridefulness, it is hardly the only source of such a warning; if it teaches us that this too shall pass, we likewise need not learn that lesson from him. If Darwin links us to animals in our origins, that is a more distant link than the one a dispassionate observer of human things like Aesop can readily see in our actions.

It seems true that Darwinism confounds both pagan and Biblical stories of human origins, and it is impossible to deny the corrosive effect Darwinism has had on Biblical faiths. Yet here again its limited impact is worth noting. The secular view that Darwinism has been taken to advance has not achieved its ultimate victory, not only because (for better and worse) faith can trump materialistic rationalism, but because faith has adapted. Rather than say Darwinism confounds the Biblical stories, it would be truer to say that Darwinism confounds certain ways of understanding those stories, and long before Darwin there were ways of reading them that did not treat them simply as scientific, historical, or journalistic accounts of events.

The transhumanist case for modifying our Bruegel-like understanding of the everyday world depends on making a very un-Darwinian move: transforming evolution from a natural and long-term process into a human project today. Is it truer to Darwin to modify his ideas in this fundamental way, or simply to let evolution take care of itself, and continue, as a great many people seem to do, to live out their lives without a concern for the ultimate fate of *Homo sapiens* and our evolutionary successors?

No, if there is a flaw in Bruegel's portrait of the everyday, it is not in how it treats Malthusian and Darwinian realities, but that it contains so little hint beyond the scattered feathers of Icarus' wings of what Heisig reminds us of—that terrible things can become mundane. But Heisig and Bruegel might agree at least on this: what is terrible in the tale of Daedalus and Icarus reflects flaws and limits that make the human story what it is. We cannot simply wish these flaws away, lifting ourselves into some new state of affairs where they will make no difference. We might hope that our innovations always be motivated by a wish to be better, but they will certainly always arise within a framework in which that wish has not yet been fulfilled, and that fact will always make their actual result uncertain.

THE POWER OF THE SEEMINGLY MUNDANE

Bruegel's painting suggests how transhumanists slight the power of the everyday, instead projecting our hopes and fears for the future onto what is essentially a blank canvas. Hoping to overcome the merely human, they look at the present from the point of view of their projection—judging the world around us as though they already understand the future—in order to give meaning and direction to present human activities. So what is important to them are the real or imagined innovations that serve as a prelude to this future whose own meaning will be beyond us. The prosthetic hand that could serve the disabled veteran is immediately attached to the pioneer who wants a third hand, and since he *wants* it, no further thought about the context in which such a thing might happen is considered necessary. If the transhumanists bother to look at anything in the past or present at all, it is only the as-yet-unrealized dreams of things like immortality or super powers. The godlike capacities that have long been wished for, and yet traditionally have been regarded as at least as much curse as blessing, if not far more curse, are turned into unambiguously normative aspirations. Such wishes become human essentials rather than aberrations.

With this blank canvas as their starting point, it does not seem so strange to transhumanists when they go on to assert that the meaning for life today, and a direction for future "progress," is to be found in an incomprehensible future. Yet that argument creates a powerful bias, a tunnel vision that focuses on developments and possibilities that make the least sense from the point of view of where we are today and for that very reason suggests they are the most important things in the world of tomorrow. At the very least, this kind of tunnel vision is not the only way of treating the future. Like Bruegel, we can admit the desirability of innovation and still value the continuities that for better and worse influence the meaning that those innovations will come to have for our lives, projecting the past and present into the future rather than the other way around.

Bruegel's alternative is not his alone; for example, a similar way of looking at the world is found in Tolstoy's *War and Peace*. Tolstoy describes how Pierre Bezukhov was changed by his experiences as a prisoner of Napoleon's army. Pierre grew up and, abandoning a way of looking at the world that sounds rather like the transhumanist vision, adopted a perspective like Bruegel's:

> All his life he had looked off somewhere, over the heads of the people around him, yet there was no need to strain his eyes, but only to look right in front of him.
>
> Formerly he had been unable to see the great, the unfathomable and infinite, in anything. He had only sensed that it must be somewhere and had sought for it. In all that was close and comprehensible, he had seen only the limited, the petty, the humdrum, the meaningless. He had armed himself with a mental spyglass and gazed into the distance, where the petty and humdrum, disappearing in the distant mist, had seemed to him great and infinite, only because it was not clearly visible. . . . Now he had learned to see the great, the eternal, and the infinite in everything, and therefore, in order to see it, to enjoy contemplating it, he had naturally abandoned the spyglass he had been looking through until then over people's heads, and joyfully contemplated the ever-changing, ever-great, unfathomable, and infinite life around him.[13]

For the advocates of the eclipse of man, the "spyglass" may be some distant future, aliens from the stars come to save us, the ever-rising trend line that brings the actual ever closer to the limits of the possible, or the posthuman Singularity. From all these imagined points of view there is little to be said for humanity as we see it in front of us—it is indeed petty, humdrum, and meaningless. Yet even if the futures they look forward to are possible, and even if there are powerful forces at work behind the innovations that would bring them about, there is no necessity to look at them through the "spyglass" that the advocates of dehumanization use. The peculiar farsightedness of the spyglass makes small, speculative things look big and important while turning things that are close up into a blur. Instead of looking "over the head" of humanity to the alien or posthuman, we can attempt to see what is right in front of us, to meet human life face to face, and at the very least not abandon it until we are certain we have understood it and appreciated it on its own terms.

G. K. Chesterton writes of a reformer who sees a gate in the road and says, "I don't see the use of this; let us clear it away." A more intelligent reformer, Chesterton says, would respond this way: "If you don't see the use of it, I certainly won't let you clear it away. Go away and think. Then, when you can come back and tell me that you *do* see the use of it, I may allow you to destroy it."[14] It is far from clear that the advocates of dehumanization have given much thought to the "use" of human beings, beyond various attempts at material explanations of why we are the way we are. There are a few exceptions: Winwood Reade and Nikolai Fedorov sought to place their arguments within an understanding of human things wider than the horizon of the power of technological possibilities. But most advocates of dehumanization, including today's transhumanists, are far more likely not only to take the world to revolve around the actual technological and scientific infrastructure, but to assume it revolves around what *might* become possible if only we clear away all old gates.

If we appreciate instead how important the seemingly mundane is in shaping our expectations and hopes about the world, we are more likely to ask why we have this or that technology, and why we want it, or why we want something different. What good does it do, or would we expect it to do? How does it fit with our vision of what makes for a good life—which of course may or may not be a vision of a good life simply.

Such questions will not be easy to answer. Since science and technology as such offer few if any resources for ethical reflection, on their own they leave us with a painfully thin understanding of the shape of human life. Furthermore, the advocates of dehumanization are just one part of a larger picture, and in our time that larger picture in commerce and the arts, in the humanities, in the natural and social sciences, is one that *often* simplifies, if not outright debases, our self-understanding in ways that reinforce the eclipse of man. If we lack the general intellectual tools required to make distinctions between progress and change, for example, or between freedom and willfulness, some of the transhumanists' heaviest lifting has already been done, for then it is easy to transform the possible into the necessary simply because it satisfies someone's desire.

THE PROGRESS OF HUMANITY

Beyond these particular blinders of our time and place, investigating the meaning of well-being has *never* been easy, because it requires a willingness to look at the question of the human good with care and seriousness, not taking the day-to-day for granted but not rejecting it dogmatically. That would mean neither the dogmatic acceptance nor the dogmatic rejection of the moral values of one's time and place. It would require avoiding cynicism and utopianism about human motives and possibilities. Such an investigation might yield a complex and mixed picture of what a good life is and how science and technology contribute to it. There will be grave uncertainties and honest disagreements along the way. We will likely find that even as individuals we have conflicting desires and visions of the good, not to speak of wider social and cultural disagreements. But the investigation is still worth undertaking if we want to speak meaningfully of "progress."

Putting all such challenges together, we begin to see why the problem of benevolence that has arisen in these pages looms so large. It can be hard enough to know what is genuinely benevolent when human beings are relating to human beings under the best of circumstances. When we start talking about benevolence directed to us by beings of unimaginable power and knowledge, the only intellectual experience we have is summarized in theodicy, the effort to understand God's goodness in light of the manifest evils of the world. That effort, which has occupied many a great mind over many centuries, is, shall we say, ongoing. Why it would be any easier to settle were we to start having to talk about artificial superintelligence or advanced aliens is far from clear.

A willingness to act on the basis of nearly complete moral ignorance relative to the central question of progress—the question of what would make for a better world—is really the only justification for the otherwise simplistic desire for the eclipse of man. Otherwise, we would surely want to adopt more *modest* expectations for a human future. Surely it is not as if the only future that is worth looking forward to, and working hard for, is one in which we can achieve anything we can imagine, where everything will be permitted. If it were, what we are left with is mere pride in novelty and superlatives, a constant one-upmanship of imaginative possibilities that diminish the worth of human beings as we actually know them. With no clear goal, direction, or purpose, with willful freedom of choice as the guiding light, how could it be otherwise?

To be clear, this kind of needed modesty is not that of the SETI advocates, who are happy to expose what they take to be human vanity about our place in the universe. For that is a patently false modesty; diminishing what we are is only a prelude to pride in the great expectations of what we can achieve once we meet up with aliens, a goad to take up our true task of creating limitless possibilities.

The kind of modesty that we need acknowledges that there is much that can and ought to be done to make human lives better, and that science and technology will play a major role in that effort. At the same time, it does not take for granted what we mean by "better," in light of the whole range of human strengths and weaknesses that we observe when we pay attention, like Bruegel, to the world in front of us. We are notoriously in-between beings, neither beasts nor gods as Aristotle famously put it.[15] How much confidence is appropriate, then, in our abilities to wield the great powers that are being promised to us? We can hardly afford to act on the basis of thinking that because we can imagine a day when we are without human vices, we can therefore ignore their reality when presented with technologies that could be used to help them flourish.

Anything we actually accomplish will be the product of limited and flawed creators, so the odds are that our creations will of necessity perpetuate those limits and imperfections. As we have seen repeatedly in these pages, the more we place our understanding of technological change within the constraints of the world out of which it actually arises and through which it must percolate, the more it seems likely that the result will never be as wonderful or as terrible as less-disciplined imaginations can so easily make it.[16] That hardly guarantees a good result, but if we are unwilling to take up seriously the question of what good means, and if we are too much influenced by the tunnel vision of the dehumanizers, we can hardly expect anything better.

Human beings, unlike other animals, can make deliberate choices to change what it means to be human. It may be that now we are seeing the beginnings of a real choice about being human at all. In attempting to confront the "grand vision" of the eclipse of man as such, we have seen how its advocates have made arguments against our humanity based on a painfully thin understanding of what it means to be human, and made promises that will lead to the demise of the goods sought even as they are fulfilled. Their project is neither as inevitable nor as rational as they would like to believe, and they are therefore certainly not excused from defending it on the moral grounds implicit in calling it progress: that it will actually create a *better* world.

Acknowledgments

THIS BOOK has taken longer to bring to fruition than I ever imagined it would. As a consequence, there are more people to thank than might otherwise have been the case. I only started to enhance my memory of such people by making a list after the process had already extended itself to such a degree that I fear there is every likelihood that I have forgotten some important assistance, kindness, or inspiration. To any of these people who might pick up this volume without finding the recognition due them, my sincere apologies.

It has been a deep pleasure for some years now to work in a variety of contexts, not the least of which is the Futurisms blog, with the talented editorial staff of *The New Atlantis*: Adam Keiper, Ari N. Schulman, Samuel Matlack, Brendan P. Foht, and Caitrin Nicol Keiper. They also served as my editors on this book. As usual, their thoughtful corrections and suggestions have the net result of making me seem a better writer and a more intelligent person than I am. My only excuse for complicity in this deception is that at least I am smart enough to recognize that is the outcome, and to acknowledge it herewith. Eric Cohen was the founding editor of *The New Atlantis*; he along with Peter A. Lawler at *Perspectives on Political Science* were the first to provide a print home for the writing I was doing on the arcane topic of human redesign, encouragement that came at a crucial moment. I also owe many thanks to the assiduous fact-checking efforts of several *New Atlantis* research assistants and interns: Maximilian de la Cal, Steven Fairchild, Barbara McClay, Maura McCluskey, and Galen Nicol. *New Atlantis* interns Moira E. McGrath and M. Anthony Mills helped compile the index. And thanks are due, too, to Roger Kimball, Heather Ohle, Carl W. Scarbrough, and the rest of the Encounter Books team for their contributions to unleashing this book on an unsuspecting public.

Having touched on the subject of finding a place to publish my thoughts on the topics this book covers, let me single out also Mary Nichols, who was instrumental in founding the Politics, Literature, and Film section of the American Political Science Association, and by so doing, and by the example of her own writing, encouraged me and many like me to think of literary analysis as a legitimate way to understand political things. Similarly, Stuart Kingsley's efforts to make a place for discussions of optical SETI at conventions of the International Society for Optical Engineering (SPIE) opened the door for this political scientist to put my thoughts before a community of natural scientists and reap the benefits thereof. Robby George and Brad Wilson likewise generously provided opportunities to air my ideas in the context of conferences hosted by the James Madison Program in American Ideals and Institutions at Princeton University.

In its early stages, my writing was supported by a sabbatical from Duquesne University and a generous grant from the Sarah Scaife Foundation. I note "early" to express my gratitude for the subsequent patience of executive director Michael Gleba, who never failed to show interest in how the book was going.

I can't even begin to thank my wife Leslie enough for the many things big and small she did over the years to give me the time, emotional support, and advice that allowed me to write. From discussion in the earliest days of this book to the last stages as she read galleys, her help has been indispensable. I am grateful that I can count on her energy and intelligence under all circumstances. My children Ted and Anna did their part by showing a remarkable tolerance for the needs of a father with such odd interests.

This book is less the product of specialized academic studies within a clearly defined discipline than it is of an academic life where, thanks to the tolerant attitude of colleagues at Duquesne University, I have been able to pursue the questions that intrigue me wherever they have led. But academic life is just one aspect of life, and when thinking about prospects for the human future, one can hardly help (even if unknowingly) drawing on that larger life experience. Such insights as I hope the book contains could have arisen as much from encounters walking about in my community or attending synagogue services as from the classroom, conferences, the library, the Internet. Its lineage is correspondingly complex. The list that follows encompasses people to whom I am indebted for a variety of the usual kinds of scholarly help: inspiring students and teachers (formal and informal), research assistance, paper discussants, or editorial advice. It also includes some intellectual sparring partners, sounding boards, and providers of moral support. I offer all my appreciation, without of course imputing any responsibility to them for arguments or ideas that, in some instances at least, I quite hope they do *not* agree with. So my thanks to: Steve Balch, Mark Blitz, Todd Bryfogle, Nigel Cameron, Mark Coda, Tobin Craig, Stephen L. Elkin, Robert Faulkner, Barbara Goldoftus, Kim Hendrickson, Shawn Igo, Leon R. Kass, Caroline Kelly, Syma Levine, Michael Lotze, Don Maletz, Michele Mekel, Thomas W. Merrill, Jacqueline Pfeffer Merrill, Andrew Morriss, Gary McEwan, Michael Myzak, Steve Ostro, Marty Plax, Betsy Rubin, Bill Rubin, Jeff Salmon, Adam Schulman, Tom Short, Ted Weinstein, and Stephen Wrinn. Thanks also to Wilfred M. McClay, whose 1994 book *The Masterless* included a chapter on Edward Bellamy that introduced me to the fascinating story "Dr. Heidenhoff's Process."

The last years of writing this book have taken place in the shadow of my Mom's developing "dementia consistent with Alzheimer's disease," as the more careful physicians seem to put it. I did not particularly need this lesson to teach me the fragility of human life, nor to bring home the depth of the love between my parents. On the one hand, then, it has reinforced my belief that we still have a great deal of progress to make simply towards alleviating *human* suffering and disease, and that however what we learn in the process of so doing might be related to grand schemes for redesigning ourselves, those schemes remain a distraction, and a rather ethically oblivious if not cruel one at that. Yet on the other hand, it has shown me more clearly than I would have wished the occasion for noble and beautiful deeds that open to loving and respectful souls, my Dad's foremost among them, as they provide the care that must be given her in the absence of cure. I cannot call these opportunities compensation, let alone silver lining, but they are a weighty moral fact, and not the first my folks taught me. So this book is dedicated to them.

NOTES

EPIGRAPH

Albert Einstein, "Freedom and Science," trans. James Gutmann, in *Freedom: Its Meaning*, Ruth Nanda Anshen, ed. (New York: Harcourt, Brace and Company, 1940), 382. The essay was later partially reprinted in various collections under the title "On Freedom."

INTRODUCTION

1 The differences among these schools of thought are important to those who have coined and adopted the different labels, even if outsiders may not immediately appreciate what is at stake. Yet in practice the adherents of one position or another often aim to adopt a "big tent" outlook that seeks to minimize the sectarian tendencies implicit in the different designations. So while my use of "transhumanism" in this book as a generic term necessarily means some nuance will be lost, the decision is likely to be controversial only to the extent that it may imply a single-mindedness that adherents to one position or another will object to when their own ideas seem to be slighted. But that problem is balanced by the fact that by using the term in this way I follow their own movement-building example.

2 Stewart Brand, "Purpose," *Whole Earth Catalog*, 1 (Fall 1968), 2. This phrase, originally rendered in italics with the word "are" underlined, was the first sentence of the mission statement Brand put at the beginning of the first issue of his *Whole Earth Catalog*, a short-lived publication that was influential in the counterculture movement, among futurists, and among the high-tech innovators of the 1970s and 1980s—a milieu in which many transhumanist ideas developed.

3 Thus, while transhumanist philosopher Nick Bostrom acknowledges that the technologies that allow human redesign may pose their own "existential risks," he also argues that the transformation of mankind may be a route to avoiding other such threats. For the briefest of statements to this effect, see his "Existential Risks and Future Technologies," www.futuretech.ox.ac.uk/existential-risks-and-future-technologies.

4 William Faulkner, "Banquet Speech" (speech, Stockholm, December 10, 1950), Nobelprize.org,nobelprize.org/nobel_prizes/literature/laureates/1949/faulkner-speech.html.

5 See, for example, the cover of the February 21, 2011 edition of *Time* magazine, which reads: "2045: The Year Man Becomes Immortal."

6 John Letzing, "Google Hires Famed Futurist Ray Kurzweil," Digits (*Wall Street Journal* blog), December 14, 2012, blogs.wsj.com/digits/2012/12/14/google-hires-famed-futurist-ray-kurzweil/.

7 Singularity University, "FAQ," singularityu.org/faq/.

8 The classic presentation of this idea is to be found in Mihail C. Roco and William Sims Bainbridge, eds., *Converging Technologies for Improving Human Performance* (Dordrecht, Netherlands: Kluwer Academic Publishers, 2003), wtec.org/ConvergingTechnologies/Report/NBIC_report.pdf.

9 David Levy, *Love and Sex with Robots: The Evolution of Human-Robot Relationships* (New York: HarperCollins, 2007). The widespread attention given this book is perhaps best indicated by the fact that its author made an appearance on *The Colbert Report* on January 17, 2008: thecolbertreport.cc.com/videos/ykpl7i/david-levy.

10 See, for example, Ray Kurzweil, *The Age of Spiritual Machines: When Computers Exceed Human Intelligence* (New York: Penguin, 1999).

11 See, for example, Hans Moravec, *Mind Children: The Future of Robot and Human Intelligence* (Cambridge, Mass.: Harvard University Press, 1988).

12 K. Eric Drexler, *Engines of Creation: The Coming Era of Nanotechnology* (New York: Anchor Books, 1990).

13 See, for example, Simon Young, *Designer Evolution: A Transhumanist Manifesto* (Amherst, N.Y.: Prometheus Books, 2006).

14 For nanotechnology and water purification see Lora Kolodny, "Puralytics CEO On Cleaning Water With Light, Winning The Cleantech Open," techcrunch.com/2010/11/25/cleantech-open-winners-puralytics/. For nanotechnology and solar energy see Xiangnan Dang, *et al.*, "Virus-templated Self-assembled Single-walled Carbon Nanotubes for Highly Efficient Electron Collection in Photovoltaic Devices," *Nature Nanotechnology* 6 (2011), 377–384, dx.doi.org/10.1038/nnano.2011.50.

15 P. W. Singer, *Wired for War: The Robotics Revolution and Conflict in the 21st Century* (New York: Penguin, 2009).

16 Tractate *Eruvin* 13b.

17 Charles T. Rubin, *The Green Crusade: Rethinking the Roots of Environmentalism* (New York: The Free Press, 1994), 181–184, 206–211.

18 For an excellent account of the important role our visions of the future play in thinking even about the present, see Yuval Levin, *Imagining the Future: Science and American Democracy* (New York, Encounter Books, 2008).

19 Eric Cohen, *In the Shadow of Progress: Being Human in the Age of Technology* (New York: Encounter Books, 2008), 81.

CHAPTER ONE: THE FUTURE IN THE PAST

1 Nick Bostrom, "The Transhumanist FAQ: A General Introduction," Version 2.1 (2003), 47, transhumanism.org/resources/FAQv21.pdf. See also Chris Wren, "Star Trek's Greatest Weakness," Mondolithic Sketchbook (blog), May 11, 2006, web.archive.org/web/20061015002913/http://mondosketch.blogspot.com/2006/05/star-treks-greatest-weakness.html.

2 Keith Michael Baker, "Marquis de Condorcet," in Paul Edwards, ed., *The Encyclopedia of Philosophy I* (New York: Collier Macmillan, 1967), 182–84.

3 David Williams, *Condorcet and Modernity* (Cambridge: Cambridge University Press, 2004), 3. See also William Doyle, *The Oxford French Revolution* (Oxford: Clarendon, 1989), 399.

4 Williams, *Condorcet and Modernity*, 42.

5 *Ibid.*, 43.

6 Marie Jean Antoine Nicolas de Caritat, Marquis de Condorcet, *Outlines of an Historical View of the Progress of the Human Mind* (Philadelphia: M. Cary and Co., 1796), 245, oll.libertyfund.org/titles/condorcet-outlines-of-an-historical-view-of-the-progress-of-the-human-mind.

7 *Ibid.*, 252.

8 *Ibid.*, 289.

9 *Ibid.*, 290.

10 *Ibid.*, 291.

11 *Ibid.*, 289.

12 *Ibid.*, 289–90.

13 *Ibid.*, 292.

14 Francis Bacon, *New Atlantis and The Great Instauration* (Wheeling, Illinois: Harlan Davidson, 1989), 71.

15 Thomas Robert Malthus, *An Essay on the Principle of Population, As It Affects the Future Improvement of Society, with Remarks on the Speculations of Mr. Godwin, M. Condorcet, and Other Writers* (London: J. Johnson, 1798), archive.org/stream/essayonprinciploomalt#page/n7/mode/2up. Condorcet's *Outlines* first appeared in English in 1795; Mathus's *Essay* came out three years later.

16 David N. Reznick, *The "Origin" Then and Now: An Interpretive Guide to the "Origin of Species"* (Princeton: Princeton University Press, 2010), 68.

17 The Darwin Correspondence Project, "Spotlight on a Correspondent: William Winwood Reade," Darwin and Human Nature (blog), May 20, 2011, darwin-humannature.wordpress.com/2011/05/20/spotlight-on-a-correspondent-william-winwood-reade/.

18 "One always turns back to Winwood Reade's *Martyrdom of Man* for renewal of faith." (W. E. B. Du Bois, *The World and Africa* [New York: The Viking Press, 1947], x.) Rhodes called the book "creepy," but also said "That book has made me what I am." (Princess Catherine Radziwiłł, *My Recollections* [London: Ibister

and Company, 1904], 342.) Wells considered Reade a "great and penetrating genius." (Leon Stover, "Applied Natural History: Wells vs. Huxley," in Patrick Parrinder and Christopher Rolfe, eds., *H. G. Wells under Revision: Proceedings of the International H. G. Wells Symposium* [Selinsgrove, Penn.: Susquehanna University Press, 1990], 127.) Orwell wrote, "however hasty and unbalanced" Reade's book may seem, it "shows an astonishing width of vision." (George Orwell, *The Collected Essays, Journalism and Letters of George Orwell*, Sonia Orwell and Ian Angus, eds., vol. 3, *My Country Right or Left*, 1940–1943 [New York, Harcourt Brace Jovanovich, 1968], 37.)

19 Sir Arthur Conan Doyle, *The Sign of the Four* (orig. 1890), in *The Annotated Sherlock Holmes*, William S. Baring-Gould, ed. (New York: Wings/Random House, 1992), vol. 1, 619. I say "perhaps" Holmes admires Reade because "remarkable" does not have to be a term of praise, and Reade's book may just be a way for Watson to while away an hour. There is less ambiguous praise later in *The Sign of the Four*: "'Winwood Reade is good upon the subject,' said Holmes. 'He remarks that, while the individual man is an insoluble puzzle, in the aggregate he becomes a mathematical certainty.'" (*Ibid.*, 666.)

20 Winwood Reade, *The Martyrdom of Man* (London: Kegan Paul, Trench, Trübner & Co., 1910), xlvii, archive.org/stream/martyrdomofmanooreaduoft.

21 *Ibid.*, 395–400.

22 *Ibid.*, 10.

23 Whether or not Reade was familiar with Immanuel Kant, the argument recalls the nineteenth-century German philosopher's position in works such as "Idea for a Universal History with a Cosmopolitan Intent," collected in Immanuel Kant, *Perpetual Peace and Other Essays* (Indianapolis: Hackett Publishing Co., 1983), 29–40.

24 Reade, *The Martyrdom of Man*, 10.

25 *Ibid.*, 392.

26 *Ibid.*, 410.

27 *Ibid.*, 393.

28 *Ibid.*, 522–523.

29 *Ibid.*, 443.

30 *Ibid.*, 467.

31 *Ibid.*, 512.

32 *Ibid.*, 512.

33 *Ibid.*, 464.

34 *Ibid.*, 452.

35 *Ibid.*, 505, 516ff.

36 Note that Reade also says, "Finally, men will master the forces of nature; they will become themselves architects of systems, manufacturers of worlds. . . . There is but a difference in degree between the chemist who to-day arranges forces in his laboratory so that they produce a gas, and the creator who arranges forces

so that they produce a world; between the gardener who plants a seed, and the creator who plants a nebula." So perhaps we will not always be so far below the "Power by whom the universe was made" after all. Reade, *The Martyrdom of Man*, 515, 467.

37 "Human knowledge and human power meet in one; for where the cause is not known the effect cannot be produced. Nature to be commanded must be obeyed." Francis Bacon, *The New Organon* (Aphorisms I.3), Fulton H. Anderson, ed. (Indianapolis: Bobbs-Merrill, 1960), 39.

38 Reade, *The Martyrdom of Man*, 513.

39 *Ibid.*, 511.

40 *Ibid.*, 515.

41 *Ibid.*, 514–15

42 *Ibid.*, 515.

43 *Ibid.*, 538–39.

44 *Ibid.*, 543.

45 Elisabeth Koutaissoff, "Introduction," in Nikolai Fedorovich Fedorov, *What Was Man Created For?: The Philosophy of the Common Task* (selected works), Elisabeth Koutaissoff and Marilyn Minto, trans. and eds. (Lausanne: Honeyglen Publishing/L'Age d'Homme, 1990), 16.

46 Charles Tandy and R. Michael Perry, "Nikolai Fedorovich Fedorov (1829–1903)" *Internet Encyclopedia of Philosophy*, iep.utm.edu/fedorov/.

47 Koutaissoff, "Introduction," 16.

48 Fedorov, *What Was Man Created For?*, 36.

49 *Ibid.*, 34.

50 *Ibid.*, 36.

51 *Ibid.*, 39.

52 *Ibid.*, 42, 43.

53 *Ibid.*, 53–54.

54 *Ibid.*, 54.

55 *Ibid.*, 65.

56 *Ibid.*, 56, 43–44.

57 *Ibid.*, 91.

58 *Ibid.*, 60–61, 66, 43.

59 *Ibid.*, 90. For an interesting comment on the likeness between Fedorov and the transhumanists, see Charles Stross, "Federov's [*sic*] Rapture," Charlie's Diary (blog), July 1, 2011, www.antipope.org/charlie/blog-static/2011/07/federovs-rapture.html.

60 Fedorov, *op cit.*, 88.

61 *Ibid.*, 96.

62 *Ibid.*, 97, 96.

63 *Ibid.*, 97.

64 *Ibid.*, 97.

65 *Ibid.*, 97–98.

66 *Ibid.*, 127–128.

67 Camille Flammarion, *Astronomy for Amateurs*, trans. Francis A. Welby (New York: D. Appleton and Co., 1915 [orig. 1904]), 319, archive.org/stream/astronomyforamaoowelbgoog.

68 *Ibid.*, 325–326.

69 *Ibid.*, 338.

70 *Ibid.*, 331.

71 *Ibid.*, 338.

72 *Ibid.*, emphasis added.

73 *Ibid.*, 337.

74 *Ibid.*, 340.

75 Richard Jeffery, "C. S. Lewis and the Scientists," *The Chronicle of the Oxford University C. S. Lewis Society* 2, no. 2 (May 2005), 17–19.

76 In the United Kingdom, "The Last Judgment" was published as the final piece in the book *Possible Worlds and Other Essays* in 1927. It was excluded from the American edition of that book, because the piece was instead published in the March 1927 edition of *Harper's Magazine* and as a small standalone book the same year by Harper & Brothers. (See Mark B. Adams, "Last Judgment: The Visionary Biology of J. B. S. Haldane," *Journal of the History of Biology* 33, no. 3 [Winter 2000], 462.) "The Last Judgment" is a sort of sequel to Haldane's 1923 lecture *Daedalus, or Science and the Future*, which I analyzed in Charles T. Rubin, "Daedalus and Icarus Revisited," *The New Atlantis* 8 (Spring 2005), 73–91.

77 C. S. Lewis, "On Science Fiction," in *Of Other Worlds: Essays and Stories* (San Diego: Harvest/Harcourt, 1994 [orig. 1966]), 66.

78 J. B. S. Haldane, *Possible Worlds and Other Essays* (London: Chatto and Windus, 1927), 292.

79 *Ibid.*, 295.

80 *Ibid.*

81 *Ibid.*

82 *Ibid.*

83 *Ibid.*, 302.

84 *Ibid.*, 303.

85 *Ibid.*

86 *Ibid.*

87 *Ibid.*, 309.

88 *Ibid.*

89 *Ibid.*, 310.

90 *Ibid.*

91 *Ibid.*, 311.

92 *Ibid.*, 311–321.

93 *Ibid.*, 312.

94 *Ibid.*, 310, 312.

95 Andrew Brown, *J. D. Bernal: The Sage of Science* (Oxford: Oxford University Press, 2006), 54.

96 *Ibid.*, 238–254, 158–159.

97 *Ibid.*, 341, 416, 432.

98 J. D. Bernal, *The World, the Flesh and the Devil: An Enquiry into the Future of the Three Enemies of the Rational Soul* (Bloomington: Indiana University Press, 1969), v.

99 *Ibid.*, 6, 5.

100 *Ibid.*, 15.

101 The first spacecraft to rely on solar sails was a Japanese probe given the acronym IKAROS, after Icarus, the mythological figure who perished after flying too close to the sun; he is depicted on the cover of this book and discussed at length in Chapter Five.

102 Bernal, *The World, the Flesh and the Devil*, 18, 25.

103 *Ibid.*, 27.

104 *Ibid.*, 32.

105 *Ibid.*, 35.

106 *Ibid.*, 34.

107 *Ibid.*, 34.

108 *Ibid.*, 37.

109 *Ibid.*, 36.

110 *Ibid.*, 37–38.

111 *Ibid.*, 42.

112 *Ibid.*, 38–41. Bernal paints a vivid picture of the mechanical men he envisions: "Instead of the present body structure we should have the whole framework of some very rigid material, probably not metal but one of the new fibrous substances. In shape it might well be rather a short cylinder. Inside the cylinder, and supported very carefully to prevent shock, is the brain with its nerve connections, immersed in a liquid of the nature of cerebro-spinal fluid, kept circulating over it at a uniform temperature. The brain and nerve cells are kept supplied with fresh oxygenated blood and drained of de-oxygenated blood through their arteries and veins which connect outside the cylinder to the artificial heart-lung digestive system—an elaborate, automatic contrivance.... The brain thus guaranteed continuous awareness, is connected in the anterior of the case with its immediate sense organs, the eye and the ear—which will probably retain this connection for a long time. The eyes will look into a kind of optical box which will enable them alternatively to look into periscopes projecting from the case, telescopes, microscopes and a whole range of televisual apparatus. The ear would have the corresponding microphone attachments and would still be the chief organ for wireless reception. Smell and taste organs, on the other hand, would be prolonged into connections outside the case and

would be changed into chemical tasting organs, achieving a more conscious and less purely emotional role than they have at present. . . . Attached to the brain cylinder would be its immediate motor organs, corresponding to but much more complex than, our mouth, tongue and hands. This appendage system would probably be built up like that of a crustacean which uses the same general type for antenna, jaw and limb; and they would range from delicate micro-manipulators to levers capable of exerting considerable forces, all controlled by the appropriate motor nerves. . . . The remaining organs would have a more temporary connection with the brain-case. There would be locomotor apparatus of different kinds, which could be used alternatively for slow movement, equivalent to walking, for rapid transit and for flight. On the whole, however, the locomotor organs would not be much used because the extension of the sense organs would tend to take their place. Most of these would be mere mechanisms quite apart from the body; there would be the sending parts of the television apparatus, tele-acoustic and tele-chemical organs, and tele-sensory organs of the nature of touch for determining all forms of textures. Besides these there would be various tele-motor organs for manipulating materials at great distances from the controlling mind. These extended organs would only belong in a loose sense to any particular person, or rather, they would belong only temporarily to the person who was using them and could equivalently be operated by other people."

113 *Ibid.*, 40, 41.

114 *Ibid.*, 41.

115 *Ibid.*, 43.

116 *Ibid.*, 46.

117 *Ibid.*, 47.

118 *Ibid.*, vi.

119 *Ibid.*, 56.

120 *Ibid.*, 53.

121 *Ibid.*

122 *Ibid.*, 56, 57.

123 *Ibid.*, 58.

124 *Ibid.*, 60.

125 Interestingly, the word "cyborg" (cybernetic organism), now popularly used to describe beings that are part living and part machine, was first coined in 1960 to propose a way to adapt human beings for the hostile environments of space exploration. See Manfred E. Clynes and Nathan S. Kline, "Cyborgs and space," *Astronautics* 14, no. 9 (September 1960), 26–27, 74–76.

126 Bernal, *The World, the Flesh and the Devil*, 63.

127 *Ibid.*, 64.

128 *Ibid.*, 65.

129 *Ibid.*, 66.

130 *Ibid.*
131 *Ibid.*, 70, 75.
132 *Ibid.*, 74.
133 *Ibid.*
134 *Ibid.*, 78–79.
135 *Ibid.*, 79.
136 *Ibid.*, 80.
137 Bernal worries that his vision may fail to be convincing for two kinds of reasons. On the one hand, perhaps he has not portrayed a future that is sufficiently "mysterious and full of supernatural power." Paradoxically, the very desire for such a future has driven the creation of the world today, a world in which we have the non-mysterious power to fulfill all our desires except this one. Will the fact that the future is now becoming "a function of our own action" turn us away from further achievement, or will we accept that an unfulfillable desire will drive us ever onwards? On the other hand, Bernal acknowledges that "we shall have very sane reactionaries at all periods warning us to remain in the natural and primitive state of humanity, which is usually the last stage but one in their cultural history." But he is not terribly worried that such critics, however sane, might actually prevail. He suggests that "the secondary consequences of what men have already done—the reactionaries as much as any—will carry them away then as now." As even reactionaries are only committed to preserving "the last stage but one," they are forced to be progressive by the brute facts that are being established around them by those who have no corresponding qualms. *Ibid.*, 80–81, 65.
138 *Ibid.*, 68.
139 *Ibid.*, 64.
140 David Pearce, "The Hedonistic Imperative" (online publication, 1995), hedweb.com/hedethic/hedonist.htm.
141 "The need to refuel three times daily is most frustrating. . . . The requirement for the daily evacuation of unpleasantly odorous waste products from an orifice needlessly situated directly next to the sexual organs is an obvious design fault." Simon Young, *Designer Evolution: A Transhumanist Manifesto* (Amherst, N.Y.: Prometheus Books, 2006), 28.
142 Jean-Jacques Rousseau, *The First and Second Discourses*, Roger D. Masters, ed. (New York: St. Martin's Press, 1964), 114–115.
143 Condorcet, 290.
144 The power of this vision of space as the new New World is such that still today, when we have gained so much knowledge about the actual technical challenges of space travel, visionary proponents of this solution to Malthus's problem remain. See, for example, Marshall T. Savage, *The Millennial Project: Colonizing the Galaxy in Eight Easy Steps* (Boston: Little, Brown and Company, 1994).

145 Timothy Shanahan, *The Evolution of Darwinism: Selection, Adaptation, and Progress in Evolutionary Biology* (Cambridge: Cambridge University Press, 2004), 173–195.

146 *Ibid.*, 276.

147 Flammarion, 325.

148 Flammarion, 325.

149 George Santayana, *The Life of Reason or The Phases of Human Progress* (New York: Charles Scribner's Sons, 1906), 284.

150 Adam Keiper, "The Age of Neuroelectronics," *The New Atlantis* 11 (Winter 2006), 4–41.

151 See the "Wi-Fi Detector Shirt" sold by ThinkGeek, thinkgeek.com/tshirts/illuminated/991e/.

152 See the account of actual experiments involving head transplants and monkey and dog brains in a bottle in Mary Roach, *Stiff: The Curious Lives of Human Cadavers* (New York: W. W. Norton, 2003).

CHAPTER TWO: DISCOVERING INHUMANITY

1 Mark Chartrand, "What To Do When ET Comes," *Ad Astra* 7, no. 2 (March/April 1995), 54.

2 Shawn Carlson, "Science and Society: SETI on Earth," *The Humanist* 51 (September/October 1991), 37.

3 Donald Tarter, "Practicing Safe Science: Treading on the Edge," *Skeptical Inquirer* 17, no. 3 (Spring 1993), 295.

4 Carl Sagan, "The Search for Signals From Space," *Parade*, September 19, 1993, 4–7.

5 Jeffrey Bennett, *Beyond UFOs: The Search for Extraterrestrial Life and Its Astonishing Implications for Our Future* (Princeton: Princeton University Press, 2008), 204.

6 Inspiration for the ending of this prologue is owed to Walker Percy, *Lost in the Cosmos: The Last Self-Help Book* (New York: Washington Square Press, 1983), 255–256.

7 Ray Kurzweil, *The Singularity Is Near: When Humans Transcend Biology* (New York: Viking: 2005), 357.

8 Nick Bostrom, "Where Are They? Why I Hope the Search for Extraterrestrial Life Finds Nothing," *Technology Review* (May/June, 2008), 72–77; quoted from nickbostrom.com/extraterrestrial.pdf, 9.

9 Eric M. Jones, "Estimates of Expansion Timescales," in B. Zuckerman and M. H. Hart, eds., *Extraterrestrials: Where Are They?* 2nd ed. (Cambridge: Cambridge University Press, 1995), 92–102.

10 Paul Davies, *The Eerie Silence: Renewing Our Search for Alien Intelligence* (Boston: Mariner Books, 2010), 167.

11 "Fermi Paradox," Wikipedia, en.wikipedia.org/wiki/Fermi_paradox.

12 This topic is sometimes addressed by scientists interested in the search for extraterrestrial intelligence (SETI). Perhaps conflating technological prowess with moral decency, pioneering SETI astronomer Jill Tarter suggests that, "If aliens were able to visit Earth, that would mean they would have technological capabilities sophisticated enough not to need slaves, food or other planets" ("Sci-Fi Movies Are Wrong About Aliens, E.T. Hunter Jill Tarter Says," Space.com, May 25, 2012, space.com/15870-aliens-earth-science-fiction-tarter-seti.html). On the other hand, her SETI colleague Seth Shostak believes that any aliens who come visit us will probably have "a more aggressive personality. And if they have the technology to come here, the idea that we can take them on is like Napoleon taking on [the] U.S. Air Force. We're not going to be able to defend ourselves very well" (Sue Karlin, "The Aliens Would Win," *IEEE Spectrum*, June 6, 2012, spectrum.ieee.org/tech-talk/geek-life/hands-on/the-aliens-would-win). The theoretical physicist Stephen Hawking periodically makes a splash by suggesting that contact with aliens might not be a good idea. For instance, in 2010 he said, "If aliens ever visit us, I think the outcome would be much as when Christopher Columbus first landed in America, which didn't turn out very well for the Native Americans. . . . I imagine they might exist in massive ships, having used up all the resources from their home planet. Such advanced aliens would perhaps become nomads, looking to conquer and colonize whatever planets they can reach" (Chris Matyszczyk, "Stephen Hawking: Aliens Might Hate Us," CNET, April 25, 2010, cnet.com/8301-17852_3-20003358-71.html).

13 Seth Shostak, *Sharing the Universe: Perspectives on Extraterrestrial Life* (Berkeley: Berkeley Hills Books, 1998), 7.

14 *Ibid.*, 190.

15 *Ibid.*

16 I. S. Shklovskii and Carl Sagan, *Intelligent Life in the Universe* (San Francisco: Holden-Day, 1966), 358.

17 Frank Drake and Dava Sobel, *Is Anyone Out There? The Scientific Search for Extraterrestrial Intelligence* (New York: Dell Publishing, 1992), 159.

18 Carl Sagan, *The Cosmic Connection: An Extraterrestrial Perspective* (New York: Anchor Books, 1973), 218.

19 Gerald Feinberg and Robert Shapiro, *Life Beyond Earth: The Intelligent Earthling's Guide to Life in the Universe* (New York: William Morrow and Co., 1980), 436.

20 Shklovskii and Sagan, *Intelligent Life in the Universe*, 412.

21 "Interview with Dr. Carl Sagan, Planetary Society Founder, SETI Pioneer" in Frank White, *The SETI Factor: How the Search for Extraterrestrial Intelligence is Changing Our View of the Universe and Ourselves* (New York, Walker and Co., 1990), 196.

22 White, *The SETI Factor*, 85.

23 Jared Diamond, "Alone in a Crowded Universe," in Zuckerman and Hart, *Extraterrestrials*, 163.

24 Zuckerman and Hart, *Extraterrestrials*, x.

25 Robin Hanson, "The Great Filter: Are We Almost Past It?" (online article), September 15, 1998, hanson.gmu.edu/greatfilter.html.

26 Sagan, *The Cosmic Connection*, 232. Somewhat more charitably, Sagan also says, "I think it is a great conceit, the idea of the present Earth establishing radio contact and becoming a member of a galactic federation—something like a bluejay or an armadillo applying to the United Nations" (242).

27 *Ibid.*, 224.

28 *Ibid.*, 219.

29 *Ibid.*

30 The English scientist William Whewell imagined unintelligent and distinctly non-humanoid alien life, based on the physics and chemistry of other planets, in his 1853 book *Of the Plurality of Worlds*; the astronomer Camille Flammarion, as we saw in Chapter One, imagined greatly varied alien life; the two French-Belgian science and science fiction writers who shared the pseudonym J. H. Rosny imagined intelligent alien life that was very different from terrestrial life; the evolutionary biologist Alfred Russel Wallace wrote in 1904 that alien life "with a mind and spiritual nature equal to that of man might have been developed in a very different form"; the naturalist Loren Eiseley, in a 1953 essay later collected in his book *The Immense Journey* (1957), argued that alien life "will not wear the shape of man"—a point on which most scientific theorists now seem to agree.

31 Nicholas Rescher, *Finitude: A Study of Cognitive Limits and Limitations* (Frankfurt: Ontos Verlag, 2011), 20–27.

32 Giuseppi Cocconi and Philip Morrison, "Searching for Interstellar Communications," *Nature* 184, no. 4690 (September 19, 1959), 844–846, dx.doi.org/10.1038/184844a0.

33 Carl Sagan, ed., *Communication with Extraterrestrial Intelligence* (Cambridge, Mass.: MIT Press, 1973), 337.

34 *Ibid.*, 337–338.

35 *Ibid.*, 337.

36 Richard Berendzen, ed., *Life Beyond Earth and the Mind of Man* (Washington, D.C., National Aeronautics and Space Administration, 1973), 44. Later collected in Philip Morrison, *Nothing Is Too Wonderful to Be True* (Woodbury, New York: AIP Press, 1995), 178.

37 Shklovskii and Sagan, *Intelligent Life in the Universe*, 396–397.

38 Drake and Sobel, *Is Anyone Out There?*, 159.

39 Greg Easterbrook, "Are We Alone?," *The Atlantic Monthly* 262, no. 2 (August 1988), 37, theatlantic.com/past/docs/issues/88aug/easterbr.htm.

40 Drake and Sobel, *Is Anyone Out There?*, 160. It should be no surprise that hopes

for immortality provide one of the themes that link SETI with technologies promoting transhuman transformations. Robert Freitas, for example, has written on xenobiology, the theoretical study of alien biology, but he is also one of the great proponents of the medical promise of nanotechnology, a source of many transhumanist hopes. (See Robert A. Freitas Jr., *Xenology: An Introduction to the Scientific Study of Extraterrestrial Life, Intelligence, and Civilization* [Sacramento: Xenology Research Institute, 1979] and his website xenology.info, as well as his multivolume work *Nanomedicine*, published by Landes Bioscience beginning in 1999, nanomedicine.com.) For an intellectual history of such links that is intelligent, illuminating, and amusing, see Ed Regis, *Great Mambo Chicken and the Transhuman Condition: Science Slightly Over the Edge* (Reading, Mass.: Perseus Books, 1990).

41 For a popular treatment of the contingent factors that shape technological development, see James Burke, *Connections* (Boston: Little Brown and Co., 1978).

42 Paul Davies, *Are We Alone? Philosophical Implications of the Discovery of Extraterrestrial Life* (New York: Basic Books, 1995), 55.

43 See, for instance, C. S. Lewis, "Religion and Rocketry," in *The World's Last Night and Other Essays* (New York: Harcourt, Brace and Company, 1960), 83–92.

44 J. B. S. Haldane, "The Last Judgment," in *Possible Worlds* (London: Chatto and Windus, 1927), 288.

45 Berendzen, *Life Beyond Earth and the Mind of Man*, 21.

46 Ray Kurzweil, *The Singularity Is Near: When Humans Transcend Biology* (New York: Viking, 2005), 21, 364–365. On this theme see Charles Stross, *Accelerando* (New York: Ace Books, 2005). Interest in searching for signs of SETI based on the more exotic possibilities that transhumanism has suggested for the human future has only grown in recent years. It is an important theme in Davies, *The Eerie Silence*, 93–115.

47 J. A. Ball, "The Zoo Hypothesis" in Donald Goldsmith, ed., *The Quest for Extraterrestrial Life: A Book of Readings* (Mill Valley, Cal.: University Science Books, 1980), 242; P. A. Sturrock, "Uncertainty in Estimates of the Number of Extraterrestrial Civilizations" in M. D. Papagiannis, ed., *Strategies for the Search for Life in the Universe* (Boston: D. Seidel, 1980), 71.

48 Sagan, *Cosmic Connection*, 222; R. N. Bracewell, "Communications from Superior Galactic Communities," in Goldsmith, *The Quest for Extraterrestrial Life*, 105–107.

49 Michael H. Hart, "An Explanation for the Absence of Extraterrestrials on Earth," in Zuckerman and Hart, *Extraterrestrials*, 1–4. But see also Sebastian von Hoerner, "Where is Everybody?" and "The General Limits of Space Travel," in Goldsmith, *The Quest for Extraterrestrial Life*, 250–54 and 197–204.

50 Sebastian von Hoerner, "Where is Everybody?" in Goldsmith, *The Quest for Extraterrestrial Life*, 250–254. Although von Hoerner notes this possibility, he does not find it convincing.

51 Davies, *The Eerie Silence*, 82.

52 *Ibid.*, 167.

53 Freeman J. Dyson, "Search for Artificial Stellar Sources of Infrared Radiation," *Science* 131, no. 3414 (June 3, 1960), 1667–1668, dx.doi.org/10.1126/science.131.3414.1667, also collected in Goldsmith, *The Quest for Extraterrestrial Life*, 108–109.

54 M. D. Papagiannis, "The Number N of Galactic Civilizations Must Be Either Very Large or Very Small," in Papagiannis, *Strategies for the Search for Life in the Universe*, 55. Cf. also Robert Lemos, "Rocket Scientists Say We'll Never Reach the Stars," Wired.com, August 19, 2008, archive.wired.com/science/space/news/2008/08/space_limits and Valerie Jamieson, "Starship Pilots: Speed Kills, Especially Warp Speed," NewScientist.com, February 17, 2010, newscientist.com/article/dn18532.

55 Arwen Nicholson and Duncan Forgan, "Slingshot dynamics for self-replicating probes and the effect on exploration timescales," *International Journal of Astrobiology* 12, iss. 4 (October 2013), 337–344, dx.doi.org/10.1017/S1473550413000244.

56 Carl Sagan, *Contact* (New York: Simon and Schuster/Pocket, 1986 [orig. 1985]), 367–368, 416–422, 430–431.

57 Sagan, *Cosmic Connection*, 222.

58 Consider, for example, the "Invitation to ETI" project founded by Allen Tough, which hopes to "establish communication with any form of extraterrestrial intelligence able to monitor our World Wide Web" (ieti.org).

59 Drake and Sobel, *Is Anyone Out There?*, 161.

60 Arthur C. Clarke, *Childhood's End* (New York: Harcourt, Brace and World, 1953), 15.

61 *Ibid.*, 16–17.

62 *Ibid.*, 24.

63 *Ibid.*, 41.

64 *Ibid.*, 42.

65 *Ibid.*, 14.

66 *Ibid.*, 58.

67 *Ibid.*, 68.

68 *Ibid.*, 72–73.

69 *Ibid.*, 136.

70 *Ibid.*, 111.

71 *Ibid.*, 135.

72 *Ibid.*, 175.

73 *Ibid.*, 209.

74 Not for nothing did C. S. Lewis say that he was "bowled over" by Clarke's novel, finding it "quite out of range of the common space-and-time writers," an "ABSOLUTE CORKER" that even "brought tears to my eyes." C. S. Lewis to Joy

Gresham, December 22, 1953, in *The Collected Letters of C. S. Lewis, Volume 3: Narnia, Cambridge, and Joy, 1950–1963*, ed. Walter Hooper (New York: HarperCollins, 2007), 390–392.

75 *Childhood's End*, 12.
76 *Ibid.*, 14.
77 *Ibid.*
78 *Ibid.*, 21.
79 *Ibid.*
80 *Ibid.*, 20.
81 *Ibid.*, 182–183.
82 *Ibid.*, 70, 74.
83 *Ibid.*, 136.
84 *Ibid.*, 95.
85 *Ibid.*, 190.
86 *Ibid.*, 77–78.
87 *Ibid.*, 90.
88 *Ibid.*, 91, 141.
89 As he describes New Athens, Clarke shows he has in mind Francis Bacon, one of the founders of modern science. The character who is "the driving force behind New Athens" is given the name Ben Salomon. In Bacon's famous story "New Atlantis," he depicts a secretive island that is home to "Salomon's House," what we would now recognize as an institute for scientific research, founded by the late King Solamona. While, in Plato's original myth, Athens was victorious over Atlantis, in Clarke's story the material satisfactions prefigured in Bacon's "New Atlantis" show their power over the creative efforts of his New Athens. (See Francis Bacon, *New Atlantis and The Great Instauration*, ed. Jerry Weinberger [Wheeling, Il.: Harlan Davidson (1989, revised edition)], 56, 58.)
90 Clarke, *Childhood's End*, 164.
91 *Ibid.*, 20, 168.
92 *Ibid.*, 161.
93 *Ibid.*, 68–69.
94 *Ibid.*, 208, 184.
95 *Ibid.*, 185.
96 *Ibid.*
97 *Ibid.*, 183.
98 *Ibid.*, 207.
99 *Ibid.*, 204.
100 *Ibid.*, 186.
101 *Ibid.*, 135.
102 *Ibid.*, 184.
103 *Ibid.*, 184.
104 That the Overmind begins to look like a stand-in for a rather traditional God is

also suggested by the fact that its angels seem to be slowly and quietly preparing a revolt. The Overlords are not quite so resigned to their task as they make themselves out to be. They use Jan Rodricks to gather information about the Overmind, both on their own planet and when the last moments of Earth approach. As a human, he can see things that are hidden from them because of their complete lack of paranormal powers. Their effort parallels the investigations that Stormgren undertakes at the end of his career. In neither case ought we to conclude that there are immediate plans or even hopes of altering the hierarchy—but if ever an effort were to be made, it would have to be based on such intelligence. The Overlords are loyal, because so far they have no choice but to obey. But they are also patient and long lived. That they want to understand their master better may be more than idle curiosity; perhaps they would like to have a choice whether or not to obey.

105 See, for example, the justification that Plutarch offers for the Spartan policies of the community of wives and the exposure of ill-formed infants in his "Lycurgus." Bernadette Perrin, trans., *Plutarch's Lives* (Cambridge, Mass.: Harvard University Press, 1914), 251–55.

CHAPTER THREE: ENABLING INHUMANITY

1 Nick Bostrom, "The Transhumanist FAQ: A General Introduction," Version 2.1 (2003), 9, transhumanism.org/resources/FAQv21.pdf.

2 Simon Young, *Designer Evolution: A Transhumanist Manifesto* (Amherst, N.Y.: Prometheus Books, 2006), 52.

3 Bostrom, "The Transhumanist FAQ," 49, 26.

4 Ray Kurzweil, *The Age of Spiritual Machines: When Computers Exceed Human Intelligence* (New York: Penguin, 1999), 145.

5 *Ibid.*, 146.

6 The Project on Emerging Nanotechnologies at the Woodrow Wilson Center has since 2005 published a "Nanotechnology Consumer Products Inventory" that, as of its October 2013 update, lists some 1,628 products that use—or at least *claim* to use—nanoparticles or nanomaterials. See nanotechproject.org/cpi/.

7 See, for example, "Engineers Build World's Smallest, Fastest Nanomotor" (press release), University of Texas at Austin, Cockrell School of Engineering, May 20, 2014, engr.utexas.edu/features/nanomotors.

8 Ortwin Renn and Mihail C. Roco, "Nanotechnology and the Need for Risk Governance," *Journal of Nanoparticle Research* 8, iss. 2 (April 2006), 153–191, dx.doi.org/10.1007/s11051-006-9092-7.

9 K. Eric Drexler, "Molecular engineering: An approach to the development of general capabilities for molecular manipulation," *Proceedings of the National Academy of Sciences* 78, no. 9 (September 1981), 5275–5278.

10 K. Eric Drexler, *Engines of Creation: The Coming Era of Nanotechnology* (New

York: Anchor Books, 1990 [orig. 1986]), 5. Drexler has placed the complete text of the book online for free at e-drexler.com/d/06/00/EOC/EOC_Table_of_Contents.html.

11 *Ibid.*, 53–63.

12 *Ibid.*, 95.

13 *Ibid.*, 81.

14 *Ibid.*, 93–95.

15 *Ibid.*, 99–116.

16 *Ibid.*, 106.

17 *Ibid.*, 130–146.

18 *Ibid.*, 111.

19 The physicist Richard Feynman is often credited as an important inspiration for nanotechnology; see Richard P. Feynman, "There's Plenty of Room at the Bottom," *Engineering and Science* 23, no. 5 (February 1960), 22–36, calteches.library.caltech.edu/1976/1/1960Bottom.pdf. For a thoughtful and critical assessment of this claim and Drexler's role, see Adam Keiper, "The Nanotechnology Revolution," *The New Atlantis* 2 (Summer 2003), 17–34, and Adam Keiper, "Feynman and the Futurists," *Wall Street Journal*, January 8, 2010, wsj.com/news/articles/SB10001424052748703580904574638160601840456.

20 National Nanotechnology Initiative, "Nanotechnology Timeline" (undated), nano.gov/timeline.

21 Drexler has written thoughtfully about how his kind of scientifically informed speculation about technological possibilities differs from the ways we usually conceive of science. His preferred term is "exploratory engineering," which emphasizes both the theoretical nature of his ideas and the fact that they are aimed at accomplishing potential technical goals. It is a mistake, Drexler argues, to judge exercises in exploratory engineering by the standards of either science or engineering, which have different goals and methods. K. Eric Drexler, *Radical Abundance: How a Revolution in Nanotechnology Will Change Civilization* (New York: PublicAffairs, 2013), 132–144. See also K. Eric Drexler, *Nanosystems: Molecular Machinery, Manufacturing, and Computation* (New York: John Wiley & Sons, 1992), 489–506, where he makes the same argument but prefers to use the term "theoretical applied science."

22 Ed Regis, "The Incredible Shrinking Man," *Wired* (October 2004), archive.wired.com/wired/archive/12.10/drexler.html.

23 *Ibid.*, 241. This phrase, which appears in Drexler's 1990 afterword to *Engines of Creation*, appears partially in italics in his book.

24 *Ibid.*, 171.

25 *Ibid.*, 172.

26 In the years since *Engines of Creation* was first published, Drexler has taken pains to downplay the "gray goo" problem, including in the 1990 afterword to *Engines*, and in his most recent book: K. Eric Drexler, *Radical Abundance*, 201*n*.

For more on the public reaction to the "gray goo" notion, see also W. Patrick McCray, *The Visioneers: How a Group of Elite Scientists Pursued Space Colonies, Nanotechnologies, and a Limitless Future* (Princeton: Princeton University Press, 2013), 192, 248, 251.

27 Drexler, *Engines of Creation*, 174.

28 *Ibid.*

29 *Ibid.*, 195.

30 *Ibid.*, 176.

31 *Ibid.*, 32.

32 *Ibid.*, 39. Drexler went on to co-found, and for many years was closely associated with, the Foresight Institute, an organization dedicated to studying these questions about nanotechnology.

33 *Ibid.*, 77.

34 *Ibid.*, 39.

35 *Ibid.*, 189.

36 *Ibid.*, 26.

37 *Ibid.*, 38. Drexler has since given further thought to the role of design versus evolution in understanding how nanotechnology might develop, but it is not clear that there is any fundamental shift in his belief that nanotechnology will allow deliberate design to replace random evolution. K. Eric Drexler, "Evolutionary Capacity: Why Organisms Cannot Be Like Machines," Metamodern (blog), August 2, 2009, metamodern.com/2009/08/02/contrasts-in-evolutionary-capacity/.

38 Drexler, *Engines of Creation*, 32.

39 *Ibid.*, 8.

40 *Ibid.*, 21.

41 *Ibid.*, 37, 103.

42 *Ibid.*

43 *Ibid.*, 76.

44 *Ibid.*, 103.

45 *Ibid.*, 102.

46 *Ibid.*, 110.

47 See, for example, Tzvetan Todorov, *Imperfect Garden: The Legacy of Humanism* (Princeton: Princeton University Press, 2002) and Marilynne Robinson, *Absence of Mind: The Dispelling of Inwardness from the Modern Myth of the Self* (New Haven, Conn.: Yale University Press: 2010).

48 Drexler, *Engines of Creation*, 21.

49 *Ibid.*, 234, 145.

50 *Ibid.*, 234–237. As Drexler sums up, "In short, we have a chance at a future with room enough for many worlds and many choices, and with time enough to explore them. A tamed technology can stretch our limits. . . . In an open future

of wealth, room, and diversity, groups will be free to form almost any society they wish, free to fail or set a shining example for the world. Unless your dreams demand that you dominate everyone else, chances are that other people will wish to share them. If so, then you and those others may choose to get together to shape a new world. If a promising start fails—if it solves too many problems or too few—then you will be able to try again. Our problem today is not to plan or build utopias but to seek a chance to try" (237).

51 *Ibid.*, 200.

52 *Ibid.*

53 *Ibid.*, 196.

54 *Ibid.*, 187–188.

55 *Ibid.*, 199.

56 *Ibid.*, 200.

57 *Ibid.*

58 *Ibid.*, 201.

59 *Ibid.*

60 *Ibid.*

61 *Ibid.*, 209–216. Drexler also endorses the idea of hypertext. Considering that he was writing in the 1980s, before the flowering of the World Wide Web, he showed foresight in anticipating that hypertext had the potential to be a Gutenberg-scale revolution in information dissemination. See 217–230.

62 *Ibid.*, 201.

63 *Ibid.*, 201. This idea of a dedicated community of the likeminded—somewhat reminiscent of the Earthlings in J. B. S. Haldane's story "The Last Judgment" who worked so hard to salvage something from the coming disaster—is one we will return to shortly in the context of Neal Stephenson's novel *The Diamond Age.*

64 *Ibid.*, 232.

65 *Ibid.*

66 *Ibid.*, 231.

67 *Ibid.*, 239. Drexler's notion of a perilous period of transition to a future stage of great power recalls the idea of the "great filter" we discussed in the context of SETI in Chapter Two.

68 *Ibid.*, 200.

69 Donella H. Meadows *et al.*, *The Limits to Growth: A Report for the Club of Rome's Project on the Predicament of Mankind* (New York: New American Library, 1972).

70 Ed Regis, *Great Mambo Chicken and the Transhuman Condition: Science Slightly Over the Edge* (Reading, Mass.: Perseus Books, 1990), 116–117. See also Drexler, *Radical Abundance*, 13–16, 246, and McCray, *The Visioneers*, 149, 153.

71 Drexler, *Engines of Creation*, 163.

72 *Ibid.*, 165, 163. It should be remembered that Drexler wrote *Engines of Creation* at a time of intense public interest in the issue of overpopulation and worries

about a "population bomb." In the decades since, world population has continued to grow but the rate of growth has slowed, and in many countries the population is now shrinking or is projected to soon shrink.

73 Drexler argues that the "settled cultures" towards the interior of the wave of human (or transhuman?) space invaders he envisions will have to deal with limits more than those on the frontier, because as settled they will have more clearly defined goals, which already implies living within limits no matter what kinds of resources are available. But they will stand in contrast with the creativity of the frontier, where "standards keep changing" and therefore "this idea of limits becomes irrelevant." More broadly, he concludes that even if "brute matter" creates ultimate limits, mind has "room for endless evolution and change." *Ibid.*, 165.

74 *Ibid.*, 237.

75 Aristotle, *The Politics* (1267a), trans. Carnes Lord (Chicago: University of Chicago Press, 2013 [second edition]), 40.

76 *Ibid.*

77 For more on this theme, see P. B. Thompson, "Social Acceptance of Nanotechnology," in Mihail C. Roco and William Sims Bainbridge, eds., *Societal Implications of Nanoscience and Nanotechnology* (Washington, D.C.: National Science Foundation, 2001), 198–202, wtec.org/loyola/nano/NSET.Societal.Implications/nanosi.pdf. See also Mary Douglas and Aaron Wildavsky, *Risk and Culture: An Essay on the Selection of Technological and Environmental Dangers* (Berkeley, Cal.: University of California Press, 1982).

78 Drexler, *Engines of Creation*, 238.

79 *Ibid.*

80 *Ibid.*

81 Drexler says he points readers to *The Diamond Age* whenever he is "asked to recommend a science fiction novel about advanced nanotechnologies." Stephenson's book, Drexler writes, is "a good read, it portrays a complex and surprising world, [and] it's not saturated with nanobots." K. Eric Drexler, "Nanotechnology in Science Fiction (and *vice versa*)," Metamodern (blog), April 9, 2009, metamodern.com/2009/04/09/nanotechnology-in-science-fiction/. (Actually, nanobots—called "nanosites" in Stephenson's novel—are crucial to the plot.)

82 Neal Stephenson, *The Diamond Age: Or, A Young Lady's Illustrated Primer* (New York: Bantam Spectra, 2003 reissue [orig. 1995]), 273.

83 *Ibid.*, 38, 40.

84 *Ibid.*, 37.

85 *Ibid.*

86 *Ibid.*, 8.

87 *Ibid.*, 30.

88 Strangely, analysts of the novel seem divided on this point. Jan Berrien Berends seems to think that Stephenson wants to show that "our future is bright, *really*

bright"—"who wouldn't want to live there?"—and that Stephenson is blind to the many flaws Berends readily perceives. Not only does this mischaracterize the book's setting and tone, but Berends himself seems oblivious to several key plot points. (Jan Berrien Berends, "The Politics of Neal Stephenson's *The Diamond Age*," *New York Review of Science Fiction* 9, no. 104 [April 1997], 15.) On the other hand, Joachim Schummer seems to err in a different direction when implying that, as an example of the "cyberpunk" or "postcyberpunk" genres, *The Diamond Age* must have a "nihilistic undertone" and "focus on human alienation," and if it weren't so fascinated with technology, it would be dystopian. (Joachim Schummer, "'Societal and Ethical Implications of Nanotechnology': Meanings, Interest Groups, and Social Dynamics," *Techné* 8, no. 2 [Spring 2005], 59.) Brian Opie catches the right tone when he notes that *The Diamond Age* is "not utopian, since the societies it describes are not only not perfected but are self-consciously reproducing inherited models of cultural practice and values." (Brian Opie, "Technoscience in Societies of the Future: Nanotechnology and Culture in Neal Stephenson's Novel *The Diamond Age* [1995]" (conference paper), February 2004, www.europe.canterbury.ac.nz/conferences/tech2004/tpp/Opie_paper.pdf, 3.) N. Katherine Hayles believes the book shows how its utopian projects fail "at every level." (N. Katherine Hayles, "Is Utopia Obsolete?" *Peace Review* 14, no. 2 [June 2002], 136, dx.doi.org/10.1080/10402650220140148.)

89 Stephenson, *Diamond Age*, 53–55.
90 *Ibid.*, 172, 331.
91 *Ibid.*, 55–57.
92 *Ibid.*, 292.
93 *Ibid.*, 286.
94 *Ibid.*, 18.
95 *Ibid.*, 108.
96 *Ibid.*, 374.
97 *Ibid.*, 322.
98 A thought echoed in the epigraph for *The Diamond Age*, from Confucius: "By nature, men are nearly alike; by practice, they get to be wide apart."
99 *Ibid.*, 20–21.
100 *Ibid.*, 185.
101 *Ibid.*, 332.
102 *Ibid.*, 9, 136, 180.
103 It is striking how frequently analysts find this failure to end scarcity unexplained or inexplicable. See Berends, "The Politics of Neal Stephenson's *The Diamond Age*," 18, and Richard Rorty, "Hope and the Future," *Peace Review* 14, no. 2 (June 2002), 152, dx.doi.org/10.1080/10402650220140166.
104 *Ibid.*, 59, 333.
105 *Ibid.*, 21.

106 *Ibid.*, 84, 23, 208; 69; 80–82.

107 *Ibid.*, 217–218, 268; 284–285; 448. The reviewer Jan Berrien Berends finds Miranda's attachment to Nell mysterious, having perhaps missed that they share a history of abuse. He can only imagine that it is based on "a *huge* dose of maternal instinct (a property in which I do not, in fact, believe)." Berends also misses the hints that Hackworth has something to do with the Mouse Army's loyalty to Nell, and so he instead draws the extraordinary conclusion that the only reason they follow her is because she is white and Victorian. (Berends, "The Politics of Neal Stephenson's *The Diamond Age*," 18, 16.)

108 Stephenson, *The Diamond Age*, 486–487.

109 *Ibid.*, 42–43.

110 *Ibid.*, 313.

111 *Ibid.*, 337.

112 *Ibid.*, 433–434.

113 *Ibid.*, 434.

114 While a skeptic with respect to the Singularity, the technology commentator Jaron Lanier (inventor of the concept of shared "virtual realities") hopes and expects that the next stage for human beings is "fuller contact between minds." He imagines advanced Martians who pity us for the poverty of our abilities to connect with each other, separated as we are into "sacks of skin." (Joel Garreau, *Radical Evolution: The Promise and Peril of Enhancing our Minds, Our Bodies—and What it Means to be Human* [New York: Doubleday, 2005], 200–202.)

115 Stephenson, *The Diamond Age*, 263, 323.

116 *Ibid.*, 365.

117 *Ibid.*, 354.

118 *Ibid.*, 304.

119 *Ibid.*, 383.

120 *Ibid.*, 37, 384.

121 *Ibid.*

122 Job 5:7 (KJV).

CHAPTER FOUR: PERFECTING INHUMANITY

1 Ray Kurzweil, *The Age of Spiritual Machines: When Computers Exceed Human Intelligence* (New York: Penguin, 1999), 146–49.

2 Hans Moravec, *Mind Children: The Future of Robot and Human Intelligence* (Cambridge: Harvard University Press, 1988), 114.

3 Simon Young, *Designer Evolution: A Transhumanist Manifesto* (Amherst, NY: Prometheus Books, 2006), 28.

4 James Hughes, *Citizen Cyborg: Why Democracies Must Respond to the Redesigned Human of the Future* (Cambridge: Westview Press, 2004), 255.

5 William Saletan, "Among the Transhumanists: Cyborgs, Self-mutilators, and the Future of Our Race," *Slate*, June 4, 2006, slate.com/id/2142987/.

6 For example, Kurzweil, who is among other things interested in altering the human digestive system, notes that the nanotechnologies that will allow that to happen are already in the works but their intended applications are diagnostic and therapeutic rather than enhancement and redesign. Ray Kurzweil, *The Singularity Is Near: When Humans Transcend Biology* (New York: Viking, 2005), 303.

7 For an extended effort to provide this kind of "common sense" defense of transhumanism, see John Harris's book *Enhancing Evolution*. In a wonderful combination of populism, disingenuousness, and Seinfeldian phrasing, Harris claims, "I have no transhumanist program or agenda. I do think there are powerful moral reasons for ensuring the safety of the people and for enhancing our capacities, our health, and thence our lives. If the consequence of this is that we become transhumans, there is nothing wrong with that, but becoming transhumans is not the agenda." John Harris, *Enhancing Evolution: The Ethical Case for Making Better People* (Princeton: Princeton University Press, 2007), 38–39.

8 See, for example, The President's Council on Bioethics, *Beyond Therapy: Biotechnology and the Pursuit of Happiness* (Washington, D.C.: U.S. GPO, 2003), 13–16, thenewatlantis.com/BeyondTherapyPDF; and Ramez Naam, *More Than Human: Embracing the Promise of Biological Enhancement* (New York: Broadway Books, 2005), 5–6.

9 U.S. Food and Drug Administration (FDA), "FDA approves first retinal implant for adults with rare genetic eye disease" (press release), February 14, 2013, www.fda.gov/NewsEvents/Newsroom/PressAnnouncements/ucm339824.htm. FDA, "FDA Approves First Implantable Miniature Telescope to Improve Sight of AMD patients" (press release), July 6, 2010, www.fda.gov/NewsEvents/Newsroom/PressAnnouncements/ucm218066.htm.

10 Lucilla Cardinali, *et al.*, "Tool-Use Induces Morphological Updating of the Body Schema," *Current Biology* 19, iss. 12 (June 23, 2009), R478–79, dx.doi.org/10.1016/j.cub.2009.05.009. See also Karen Hopkin, "Tools Are Body Parts to Brain," *Scientific American* (podcast), scientificamerican.com/podcast/episode/tools-are-body-parts-to-brain-09-06-23/.

11 If we could see better in low light, we would use less energy for lighting; such an enhancement could be a way of addressing global warming. See S. Matthew Liao, Anders Sandberg, and Rebecca Roache, "Human Engineering and Climate Change," *Ethics, Policy and Environment* 15, iss. 2 (2012), dx.doi.org/10.1080/21550085.2012.685574.

12 Young, *Designer Evolution*, 286–287. Cf. Kurzweil, *The Singularity Is Near*, 338.

13 "Democratic transhumanist" James Hughes is not so happy with the thought of a purely market-based system of access to enhancements, and holds that social

justice would require "subsidies and universal provision." Hughes, *Citizen Cyborg*, 233.

14 See the brief scenario presented in Joel Garreau, *Radical Evolution: The Promise and Peril of Enhancing Our Minds, Our Bodies—and What It Means to Be Human* (New York: Doubleday, 2005), 7–8.

15 We will get to "the Singularity" shortly, but the following quote gives an idea of the crusading mentality transhumanists can adopt: "We are willing to do whatever it takes, within reason, to get a positive Singularity. Governments are not going to stop us. If one country shuts us down, we go to another country. . . . Just because you don't want it doesn't mean that we won't build it." Michael Anissimov, "Response to Charles Stross's 'Three Arguments Against the Singularity,'" Accelerating Future (blog), acceleratingfuture.com/michael/blog/2011/06/response-to-charles-stross-three-arguments-against-the-singularity/.

16 Nick Bostrom flirts with the idea that a time may come when parents have a "moral responsibility" to enhance their children. But he concludes that "Only in extreme and unusual cases might state infringement of procreative liberty be justified. If, for example, a would-be parent wished to undertake a genetic modification that would be clearly harmful to the child or would drastically curtail its options in life, then this prospective parent should be prevented by law from doing so." (Nick Bostrom, "The Transhumanist FAQ: A General Introduction," Version 2.1 (2003), 21, transhumanism.org/resources/FAQv21.pdf.) But note that it is not hard to imagine that a time may come when a failure to enhance, genetically or otherwise, would "drastically curtail" life options. James Hughes is also in this ambiguous situation given his belief that enhancement should be subsidized by governments. Under these circumstances it may indeed be true that one will not be coerced by government into being enhanced, but one will be coerced into paying for others to be enhanced.

17 See for instance Max More, "A Letter to Mother Nature," MaxMore.com (personal website), August 1999, web.archive.org/web/20130324082510/http://www.maxmore.com/mother.htm.

18 John Harris finds it odd that there should be anyone who is against enhancement; after all, does not enhancement mean "a difference for the better"? And who would not want that? Harris, *Enhancing Evolution*, 36.

19 See, for example, Harris, *Enhancing Evolution*, 38 and Kurzweil, *The Singularity Is Near*, 322.

20 Harris, *Enhancing Evolution*, 13, 16. Cf. Naam, *More Than Human*, 9.

21 Bostrom, "The Transhumanist FAQ," 36.

22 Kurzweil, *The Singularity Is Near*, 35–110, but cf. 14–21.

23 David Pearce, "The Hedonistic Imperative" (online publication, 1995), hedweb.com/hedethic/hedonist.htm.

24 Pearce, "The Hedonistic Imperative," hedweb.com/hedethic/hedon4.htm.

25 Moravec, *Mind Children*, 117. For the same idea, see also Kurzweil, *The Singularity Is Near*, 258.

26 Moravec, *Mind Children*, 108–110.

27 *Ibid.*, 107–108.

28 *Ibid.*, 115–116.

29 *Ibid.*, 121–122.

30 Moravec offers what he calls a "fun" kind of resurrection, just as he offers a kind of immortality: "imagine an immense simulator (I imagine it made out of a superdense neutron star) that can model the whole surface of the earth on an atomic scale and can run time forward and back.... Because of the great detail, this simulator models living things, including humans, in their full complexity. According to the pattern-identity position [that Moravec holds], such simulated people would be as real as you or me.... [W]e would bring people out of the simulation by ... linking their minds to an outside robot body, or uploading them directly into it. In all cases we would have the opportunity to recreate the past and to interact with it in a real and direct fashion. It might be fun to resurrect all the past inhabitants of the earth this way and give them an opportunity to share with us in the (ephemeral) immortality of transplanted minds. Resurrecting one small planet should be child's play long before our civilization has colonized even its first galaxy." *Ibid.*, 123–124.

31 Kurzweil, *The Singularity Is Near*, 310–320.

32 Nick Bostrom, "When will Computers Be Smarter Than Us?," Forbes.com, June 22, 2009, forbes.com/2009/06/18/superintelligence-humanity-oxford-opinions-contributors-artificial-intelligence-09-bostrom.html. This notion dates back at least to the 1960s, when mathematician I. J. Good wrote that "the first ultraintelligent machine is the last invention that man need ever make." Irving John Good, "Speculations Concerning the First Ultra-intelligent Machine," in Franz L. Alt and Morris Rubinoff, eds., *Advances in Computers*, vol. 6 (New York: Academic Press, 1965), 31–88. More recently, the idea has been taken as the central thesis for a book about artificial intelligence: James Barrat, *Our Final Invention: Artificial Intelligence and the End of the Human Era* (New York: Thomas Dunne Books, 2013).

33 Vernor Vinge, "The Coming Technological Singularity: How to Survive in the Post-Human Era" (1993 conference paper hosted on Vinge's faculty website), www-rohan.sdsu.edu/faculty/vinge/misc/singularity.html. See also Vinge's later annotated version of this essay: Vernor Vinge, "What Is the Singularity?," *Whole Earth Review* (Winter 1993, annotated Spring 2003), wholeearth.com/uploads/2/File/documents/technological_singularity.pdf.

34 Moravec, *Mind Children*, 116.

35 This is the idea of the "Dyson sphere" or "Dyson shell" that so often pops up in science fiction, named for the theoretical physicist who was the first to

mention it, in passing, in a scientific journal: Freeman J. Dyson, "Search for Artificial Stellar Sources of Infrared Radiation," *Science* 131, no. 3414 (June 3, 1960), 1667–1668, dx.doi.org/10.1126/science.131.3414.1667.

36 Kurzweil, *The Singularity Is Near*, 349–352. Kurzweil does note that "Dyson Shells can be designed to have no effect on existing planets, particularly those, like Earth, that harbor an ecology that needs to be protected" (350). But it is not clear that this statement is anything more than a pious sentiment, given the overall direction of his argument at this point; he has just calculated the computing power latent in the mass of the solar system as a whole, "not including the sun (which is ultimately also fair game)" (349). There is not much point in trying to protect Earth's ecosystems without a sun.

37 Moravec is a bit hazy on just why posthumans would wish to engage in wholesale resurrection efforts; as we saw above (note 30) it might be "fun" "child's play." Eric Drexler was similarly vague about his own version of resurrection involving frozen bodies. But we can at least note the interesting progression that has been made in the ideal of resurrection from Fedorov's solemn religious obligation to Moravec's idea of what might amuse hyperintelligence.

38 J. D. Bernal, *The World, the Flesh and the Devil: An Enquiry into the Future of the Three Enemies of the Rational Soul* (Bloomington: Indiana University Press, 1969), 66.

39 Kurzweil, *The Singularity Is Near*, 366.

40 *Ibid.*, 353–357.

41 Moravec, *Mind Children*, 147–148; Kurzweil, *The Age of Spiritual Machines*, 260.

42 Kurzweil, *The Singularity Is Near*, 342–349; but cf. Moravec, *Mind Children*, 136–139.

43 See, for example, Wendell Wallach and Colin Allen, *Moral Machines: Teaching Robots Right from Wrong* (Oxford: Oxford University Press, 2009); J. Storrs Hall, *Beyond AI: Creating the Conscience of the Machine* (Amherst, N.Y.: Prometheus Books, 2007); and Eliezer Yudkowsky, "Creating Friendly AI 1.0: The Analysis and Design of Benevolent Goal Architectures" (white paper), Machine Intelligence Research Institute (formerly the Singularity Institute for Artificial Intelligence), June 15, 2001, intelligence.org/files/CFAI.pdf. See also my essay, Charles T. Rubin, "Machine Morality and Human Responsibility," *The New Atlantis* 32 (Summer 2011), 58–79.

44 Kurzweil, *The Singularity Is Near*, 297–298. The nonbiological intelligence Kurzweil expects to exist by 2045 will be "one billion times more powerful than all human intelligence today" (136); by the end of the twenty-first century, he expects "the nonbiological portion of our intelligence will be trillions of trillions of times more powerful than unaided human intelligence" (9).

45 *Beyond Therapy*, 211.

46 *Ibid.*, 205–273.

47 Edward Bellamy, *Dr. Heidenhoff's Process* (New York: D. Appleton and Company, 1880), 72, archive.org/stream/drheidenhoffsproobellgoog.

48 *Ibid.*, 108.

49 *Ibid.*, 115–116, 115, 116, 114.

50 *Ibid.*, 103–105.

51 *Ibid.*, 100. It should be noted that, while brain researchers today accept that memories are sometimes stored in very specific locations in the brain, they do not believe that, as Dr. Heidenhoff puts it, "conventionally or morally morbid or objectionable" memories can bring about "a morbid state of the brain fibers."

52 *Ibid.*, 101.

53 *Ibid.*, 103.

54 *Ibid.*

55 *Ibid.*, 102.

56 *Ibid.*, 105.

57 *Ibid.*

58 *Ibid.*, 117.

59 *Ibid.*, 105.

60 *Ibid.*, 116–117.

61 *Ibid.*, 116.

62 *Ibid.*, 119.

63 *Ibid.*

64 *Ibid.*, 120.

65 *Ibid.*

66 *Ibid.*, 121.

67 *Ibid.*, 124, 121.

68 *Ibid.*, 125.

69 *Ibid.*, 126.

70 *Ibid.*

71 Charlie Kaufman, *Eternal Sunshine of the Spotless Mind: The Shooting Script* (New York: Newmarket Press, 2004), 38.

72 *Ibid.*, 107.

73 *Ibid.*, 58.

74 Mary may be the most serious character in the movie; at least, in the shooting script she has some intimation of what is genuinely interesting about the *Bartlett's Quotations* she loves so much—it represents the human race in "constant conversation with itself" (*Ibid.*, 64).

75 This insight into how forgetfulness compromises our ability to learn from mistakes suggests a deeper meaning to the fact that the happy conclusion of the Madeline and Henry story takes place only in his dream. Shorn of the experience that made her sadder and wiser, would not the innocent Madeline that is restored to Henry resume the "mirthful, self-reliant" ways that got her in

trouble the first time around? Even in Henry's dream there are signs of such a restoration in her "flashing, imperious expression." But she is spared by another *deus ex machina* of sorts: trying on her wedding dress, turning her expression to "shy and blushing softness" with "a host of virginal alarms." Perhaps marriage would tame her—and perhaps not. At any rate, that is Henry's dream for her. (Bellamy, *Dr. Heidenhoff's Process*, 136.)

76 Marcel Proust, *In the Shadow of Young Girls in Flower*, trans. James Grieve (New York: Penguin Books, 2002), 99.

77 Alexander Pope, "Eloisa to Abelard," poetryfoundation.org/poem/174158.

78 Tor Nørretranders, *The User Illusion: Cutting Consciousness Down to Size* (New York: Penguin Books, 1999).

79 Thomas Hobbes, *Leviathan* (New York: Collier Books, 1962 [orig. 1651]), 80.

80 Harris, *Enhancing Evolution*, 137.

CHAPTER FIVE: THE REAL MEANING OF PROGRESS

1 Winwood Reade, *The Martyrdom of Man* (London: Kegan Paul, Trench, Trübner & Co., 1910), 523, archive.org/stream/martyrdomofmanooreaduoft.

2 J. B. S. Haldane, *Daedalus, or Science and the Future* (New York: E. P. Dutton & Company, 1924), 88.

3 For example, when the Obama administration dissolved the Bush-era President's Council on Bioethics, which routinely attempted to pose the question of science and ethics in terms of a good human life, the relief in some circles was palpable. The promise was that this supposedly too-philosophical council would be replaced by a body that would instead provide so-called "practical" guidance. See Charles T. Rubin, "Postmodernism, Autonomy and Bioethical Boundaries," *The Good Society* 19, no. 1 (2010), 28–32, dx.doi.org/10.1353/gso.0.0090.

4 See, for example, Hans Jonas, "Technology and Responsibility: Reflections on the New Tasks of Ethics," in *Philosophical Essays: From Ancient Creed to Technological Man* (Chicago: University of Chicago Press, 1974), 19.

5 I should acknowledge openly what will in any case become evident soon enough: I do not approach these paintings as an art expert or art historian. But I believe looking at them naïvely can serve a useful purpose.

6 Alan Barnett, "Bernhard Heisig: The Sound and Fury of Painting," Politics and Art (blog), December 1, 2005, politicsandart.com/2005/12/bernhard-heisig-sound-and-fury-of_01.html.

7 *Ibid.*

8 Ovid, *The Metamorphoses of Ovid*, trans. Allen Mandelbaum (New York: Harcourt Brace and Co., 1993), 254.

9 My reading of the painting owes much to W. H. Auden, "Musée des Beaux Arts" (collected in W. H. Auden, *Selected Poems*, ed. Edward Mendelson [New York:

Vintage, 2007], 87) and William Carlos Williams, "Landscape with the Fall of Icarus" (collected in William Carlos Williams, *Collected Poems: Volume II, 1939–1962*, ed. Christopher MacGowan [New York: New Directions, 1991], 385–386).

10 Ovid, 256.

11 J. B. S. Haldane notes, "There is no great invention, from fire to flying, which has not been hailed as an insult to some god. . . . The biological invention then tends to begin as a perversion and end as a ritual supported by unquestioned beliefs and prejudices" (Haldane, *Daedalus*, 45, 49).

12 Compare Larry Arnhart, *Darwinian Conservatism* (Charlottesville, Va.: Imprint Academic, 2005).

13 Leo Tolstoy, *War and Peace*, trans. Richard Pevear and Larissa Volokhonsky (New York: Alfred A. Knopf, 2007), 1104.

14 G. K. Chesterton, *The Thing: Why I Am a Catholic* (orig. 1929), collected in *The Collected Works of G. K. Chesterton*, vol. 3 (San Francisco: Ignatius Press, 1990), 157.

15 Aristotle, *The Politics* (1253a), trans. Carnes Lord (Chicago: University of Chicago Press, 2013 [second edition]), 5.

16 For much more on this theme, see Peter Augustine Lawler, *Stuck with Virtue: The American Individual and Our Biotechnological Future* (Wilmington, Del.: ISI Books, 2005).

INDEX

A NOTE ON THE TYPE

ECLIPSE OF MAN has been set in Minion, a type designed by Robert Slimbach in 1990. An offshoot of the designer's researches during the development of Adobe Garamond, Minion hybridized the characteristics of numerous Renaissance sources into a single calligraphic hand. Unlike many early faces developed exclusively for digital typesetting, drawings for Minion were transferred to the computer early in the design phase, preserving much of the freshness of the original concept. Conceived with an eye toward overall harmony, Minion's capitals, lowercase letters, and numerals were carefully balanced to maintain a well-groomed "family" appearance—both between roman and italic and across the full range of weights. A decidedly contemporary face, Minion makes free use of the qualities Slimbach found most appealing in the types of the fifteenth and sixteenth centuries. Crisp drawing and a narrow set width make Minion an economical and easygoing book type, and even its name evokes its adaptable, affable, and almost self-effacing nature, referring as it does to a small size of type, a faithful or favored servant, and a kind of peach.

SERIES DESIGN BY CARL W. SCARBROUGH